AF072851

ANIMAL SUPER HEROES

WRITTEN BY CAMILLA DE LA BEDOYERE
ILLUSTRATED BY DAVID DEAN
FOREWORD BY DR JESS FRENCH

EDITED BY LAUREN FARNSWORTH
DESIGNED BY JANENE SPENCER
COVER DESIGN BY JOHN BIGWOOD

First published in Great Britain in 2019 by Buster Books,
an imprint of Michael O'Mara Books Limited,
9 Lion Yard, Tremadoc Road, London SW4 7NQ

This updated paperback edition published in 2024

 www.mombooks.com/buster
 Buster Books
 @BusterBooks
 @buster_books

Illustrations and layouts © Buster Books 2019, 2024
With additional illustrations from Shutterstock

All rights reserved. No part of this publication may be reproduced, stored in a retrieval system, or transmitted in any form or by any means, electronic, mechanical, photocopying, recording or otherwise, without prior permission of the publisher, nor be otherwise circulated in any form of binding or cover other than that in which it is published and without a similar condition including this condition being imposed on the subsequent purchaser.

ISBN: 978-1-83725-017-2

2 4 6 8 10 9 7 5 3 1

This book was printed in July 2024 by Leo Paper Products Ltd,
Heshan Astros Printing Limited, Xuantan Temple Industrial Zone,
Gulao Town, Heshan City, Guangdong Province, China.

ANIMAL SUPER HEROES

BUSTER BOOKS

CONTENTS

Foreword	6	Snow Dogs	40
Introduction	7	Other Snow Animals	41
Mari to the Rescue	8	What is a Beast of Burden?	42
Can Animals Predict Earthquakes?	10	Doctor Dogs to the Rescue!	43
Instincts and Intuition	11	Digit: The Gorilla in the Mist	44
Surprise Heroes	12	Human Ape Heroes	47
Koko: A Fine Animal Gorilla	14	Gorilla Danger	48
Can Animals Talk?	16	Wild Gorillas	49
Living with Humans	18	Wojtek the Soldier Bear	50
Bob the Street Cat	20	Mythical, Magical Bears	52
Heroic Cats	23	Dangerous Bears: Did You Know?	54
Cats: Did You Know?	24	Old Blue: On the Brink of Extinction	56
OR7 the Lone Wolf	26	What is Extinction?	58
The Wild Dog Family	28	Alien Invaders	59
Coming Home	30	The Last of Their Kind	60
Leaders of the Pack	31	Lin Wang Goes to War	62
Keiko the Unlucky Orca	32	Elephants: Did You Know?	64
Wild Whales: Did You Know?	34	Elephant Stories	66
Super Cetaceans	36	Lefty the Pit Bull Takes a Bullet	68
Brave Balto Beats the Blizzards	38	Dogs: Did You Know?	70

Medals for Heroes	72	Media Stars	108
Congo the Artistic Chimp	74	Nature's Lab	109
Famous Chimps	76	A Lion Called Christian	110
Animals and Art	77	The Family Cats	113
Can Animals Use Tools?	78	Born Free, Live Free	114
Primates in Danger	79	Juan Salvador Makes Friends	116
Machli the Crocodile Killer	80	Penguin Parents: Did You Know?	119
Deadly Killers	83	Penguin Life	120
Top Animal Parents	84	Oceans in Peril	121
GI Joe Saves the Day	86	Unlikely Heroes	122
How Do Animals Find their Way?	89	Quiz	124
Incredible Journeys	90	Index	126
Beau and Beatrice	92		
Heroic Horses	94		
Horses: Did You Know?	96		
Smoky Takes to the Skies	98		
Our Best Friends	101		
War Dog Stories	102		
Montauciel and Her Fowl Friends	104		
Animals in Space	106		

FOREWORD BY DR JESS FRENCH

Animals never cease to amaze me. That's the beauty of studying them – there is always something new to learn. I was delighted to read these tales of astonishing courage, unbelievable intelligence, astounding strength and boundless love, many of which I had never heard before. Isn't it incredible that gorillas and orangutans are capable of learning sign language? And that just one pair of robins could be responsible for saving their entire species?

I was particularly touched by the stories of animals that had risked their lives to save humans. We owe so much to the animals of our planet, yet we often don't treat them with the great respect they deserve. If we don't make changes soon, some of the animals featured in this book could end up disappearing altogether.

Young people like you are more aware and passionate than ever before. And I'm sure that after reading this book, you will be inspired to do something to protect them. Luckily, the power to make a positive change is in your hands. You might just be one person but, like Old Blue (a Chatham Island robin who you will find out about in this book), you have the power to save a whole species.

So please take inspiration from these amazing animals. Remember their stories and draw courage from their strength. And maybe, one day, you will find there is a superhero inside you too.

Jess French

INTRODUCTION

Not all heroes are human. They come in all sorts of shapes and sizes – they can be fluffy, scaly, feathered or fanged, swim, run, hop or fly.

This book contains 19 captivating tales of animals who have accomplished the most astonishing deeds, for their human companions, their families or even their entire species. Told are the stories of Mari the Shiba Inu who saved her owner from the destruction of an earthquake, Digit the gorilla who fearlessly protected his family against poachers and Old Blue the Chatham Island black robin who played the starring role in saving her entire species from extinction.

And the exploration doesn't stop there. The hidden lives of animals are often fascinating and it is thought by many that animals are far more intelligent than we give them credit for. Within these pages you can learn not only the true-life stories, but also about the most intriguing parts of animal life. Find out how whales work in teams, how wolf packs communicate with each other, how animals migrate across remarkably large distances and much more.

Humans are only one species of life on this incredible planet. Let this book open your eyes to the wonderful animals that live alongside us.

MARI
— TO THE —
RESCUE

On the morning of 23rd October, 2004, a dog named Mari gave birth to three puppies in Yamakoshi Village, Japan. That evening, a severe earthquake shook the village. Almost all the buildings, including Mari's owners' house, were destroyed. During the tremors, the newborn puppies were jolted away from their mother, and since their eyes were still closed, they could not find their way back to her. Desperate to help them, Mari broke free of her leash, picked up the puppies by the scruffs of their necks and moved them to safety. Then, without hesitation, she ran back into the collapsing building.

The grandfather of Mari's owners lived on the second floor. He was old and unwell, and struggled to climb the stairs without help. As the quake struck, a wardrobe had toppled, pinning him underneath. As he began slipping into unconsciousness, Mari appeared in the doorway. She looked at him with encouraging eyes and licked his face. It made him concentrate and unconsciousness receded.

Mari's paws were cut and bleeding from the pieces of glass and porcelain that lay all over the floor of the house. Occasionally, she disappeared from the grandfather's room and hurried downstairs to check on her puppies before reappearing at his side. Each trip gave her new wounds, but she returned again and again.

Mari managed to kindle hope in the trapped man. He pushed at the wardrobe with all his strength and eventually he was free. Slowly, he began climbing down the stairs. It took him two hours, but Mari encouraged him all the way. Upon reaching the ground floor, he escaped the precarious building and was overjoyed to find the three newborn puppies safe and sound outside too.

EARTHQUAKES STRIKE SUDDENLY.
Even today, scientists have no way of knowing exactly when or where the next one will hit.

So how come animals know before humans?

One theory suggests animals can feel the earth vibrate before people can. Another that they can detect electrical changes in the air or smell gases released as the ground fractures.

An estimated **500,000** earthquakes are detected in the world each year. Of those, 100,000 are felt by humans and 100 cause damage.

CAN ANIMALS PREDICT EARTHQUAKES?

Throughout history, there is evidence of animals reacting quickly to earthquakes.

- In Ancient Greece, rats, snakes and weasels deserted Helice days before an earthquake devastated the city.

- People have seen catfish acting strangely, chickens that stop laying eggs and even bees leaving their hive in a panic.

- In 2011, at a zoological park in Washington DC, USA, apes were seen climbing into trees some minutes before the shakes were noticed by the zoo staff.

On 26th December, 2004, a quake off northern Sumatra, Indonesia, caused a devastating tsunami that killed hundreds of thousands of people. However, very few wild animals were found dead. Scientists think that the animals had fled into the interior of the islands before the arrival of the waves.

HAD THEY SENSED THE QUAKE?

INSTINCTS AND INTUITION

DOGS CAN TELL when somebody is in distress. Scientists have observed dogs go to someone who is crying, and lick or nuzzle them. This happens even if the dog has never met the person before. They will even ignore their owner and go to the crying stranger instead.

Puppies are born very helpless. Their eyes are closed, their ears are sealed and they cannot walk until they are about three weeks old. Mothers move their puppies away from danger by carrying them by the loose skin on the back of their neck.

A dog's sense of smell is much more powerful than our own. Not only can they smell things more easily, but they are better at telling different smells apart. When Mari lost her puppies, she would have been able to locate each of them by their individual scent.

Dogs have lived with people for at least **32,000** years.

Before dogs were people's best friends, they were wild wolves who fought early humans for food and were a dangerous threat. Gradually, people began to tame them. Humans who needed to gather their food, began to use dogs to **HELP THEM TO HUNT.**

SURPRISE HEROES

There are many stories of heroic dogs who have saved lives, but these animals have amazed everyone by their unusual actions!

CAMELS are famous for their stubborn natures. So it's surprising that one of these desert nomads became famous for his community service and was awarded the rank of sergeant. Bert was a dromedary – the one-humped members of the camel family – and he lived in California, USA, where he worked alongside the Sheriff's Department. This friendly camel and his handler Nance Fite went into schools to help teach children about the dangers of drugs, and he visited youngsters in hospitals too. On one occasion, Sergeant Bert and Nance took some kids on a camping trip. Bert refused to stay outside the tent, so Nance let him sleep in the tent with them. The next morning she realized that mountain lions were in the area, and Sergeant Bert had wanted to protect her and the children from them.

In 2009, a **PARROT** was hailed a hero after it saved a toddler's life. Hannah, aged 2, was being cared for by a childminder in Denver, USA, when she began to choke on her food. The childminder, who was in the bathroom at the time, heard Willie the parrot squawking, flapping his wings and calling 'Mama baby!' She ran over to the child, who was turning blue, and managed to dislodge the food and, with thanks to Willie, saved Hannah's life.

No one would expect a rabbit to save a human life, but that's what Dory the **FLEMISH GIANT RABBIT** did. He was the pet of Simon Steggall, a British man who has diabetes. One day, Simon returned home from work and settled into his favourite chair, exhausted. Dory jumped up onto his lap for a cuddle and, when everything fell silent, Simon's wife Victoria thought they were both enjoying a nap. Suddenly, Dory began pummelling at Simon's chest, scratching at his shirt. Victoria realized that the rabbit was making a fuss because Simon had passed out. He had fallen into a diabetic coma – which can be deadly if not treated quickly. Victoria immediately called an ambulance and, thanks to Dory, Simon was soon on the mend.

One poor child suffered terribly when he went on a fishing trip with his parents in Ontario, Canada. The boy stayed on the shore while his parents were on a boat, and he witnessed the boat capsize and his parents drown. Terrified, the child tried to walk back to town alone, but the Sun went down before he got there. The youngster lay down on the ground, scared and in shock. As he began to fall asleep he felt a warm body next to his, and thought a friendly dog had found him. When he woke the next morning, the boy discovered that **THREE WILD BEAVERS** had snuggled up next to him. Their warm furry bodies had prevented the child from freezing to death that night.

KOKO
A FINE ANIMAL
GORILLA

In 1971, a baby western lowland gorilla was born in a San Francisco Zoo, USA. Her beginnings were not remarkable, but Koko went on to lead an extraordinary life of almost 47 years.

Penny Patterson, a young psychologist, hoped to discover if a gorilla could be taught sign language – the system of hand gestures that are used to communicate words and ideas in the deaf community. She began a long relationship with Koko and taught the young gorilla hundreds of signs. Soon Koko was able to string signs together to communicate her needs and wants, from requesting food to play time. She even developed her own unusual insults for people when she was angry with them, calling them a 'rotten-stink'. Koko called herself a 'fine animal gorilla'.

The scientists working with Koko gave her a kitten as a pet. Koko named it 'All Ball' and loved cuddling her furry friend. When All Ball died, Koko signed 'Cry, frown, sad, trouble'.

People all over the world were intrigued by this gentle giant, and were surprised to learn that gorillas are clever animals with emotions that are similar to our own. There were books written about Koko, the amazing 'talking gorilla', and many people, including celebrities, were filmed visiting her.

Today, ape-human language experiments are rarely conducted, as it is considered unethical to take them away from their families. One of Koko's greatest gifts was to teach us to respect the uniqueness of her species, and the importance of gorillas' natural habitat. There they can lead full, natural and wild lives, and communicate with their own families.

CAN ANIMALS TALK?

Animals communicate with each other all the time. The problem is we don't usually understand them, so we have no idea what they are saying! However, we are getting more skilled at recognizing some of the incredible ways animals can talk to each other.

When a **WILD DOG** rolls over onto its back it's using body language to talk to other dogs. By showing its soft belly, the dog is signalling that it knows the other dog is bigger or stronger, and doesn't want to fight it. Pet dogs do this too.

CHIMPS have very expressive faces, just like humans. They pout when they are in distress, grin when they are nervous and open their mouths wide to make a 'play face' when they want to have fun.

One of the simplest ways that animals communicate is through smell. When animals rub their bodies against trees, or spray urine against bushes, they are leaving their body smell to tell other animals to stay away. These chemical messages can also be used to tell a possible mate where to find them. Even tiny creatures, such as **ANTS**, use chemicals to communicate.

MEERKATS purr when they are in a good mood, and **GRASSHOPPERS** chirp to find mates. **COYOTES** howl to let members of their pack know where they are. **DEATH'S HEAD HAWKMOTHS** can squeak like a queen bee. They use this clever trick to fool bees into thinking they are friendly, before stealing their honey.

HONEYBEES perform a special waggle dance when they return to the hive. It tells other bees where to go to find the best flowers.

HOWLER MONKEYS are some of the world's loudest animals. At dawn their calls can be heard several kilometres away across the treetops. It's a useful way for different troops of howler monkeys to tell their neighbours where they are, so they can keep their distance and not compete for the same food.

Often believed to be one of the most intelligent breeds of dog, **BORDER COLLIES** are often used as sheepdogs as they are so good at following commands. But Chaser the border collie takes the biscuit – she can recognize the names of 1,022 toys, which means she has learnt more human words than any other animal.

Alex the Parrot

One of the world's most famous talking animals was Alex, an **AFRICAN GREY PARROT**. Dr Irene Pepperberg thought that parrots – being birds that can mimic human speech – would make good subjects for her research into language. But she wasn't prepared for what she discovered.

Alex helped her to demonstrate that parrots are surprisingly smart animals. He learned more than **150 WORDS** that he could both say and understand. He was even able to understand words with complex meanings, such as 'different', 'same', 'larger' and 'smaller'. Alex could use number words to count up to six objects and he could also name five different shapes and seven colours.

LIVING WITH HUMANS

Humans are members of the ape family, and share many similarities not just in the way they look but also in the way they behave. It's no wonder that in the past people have attempted to include apes and monkeys in their lives, but these experiments have often met with failure because most wild animals are much better off with their own families.

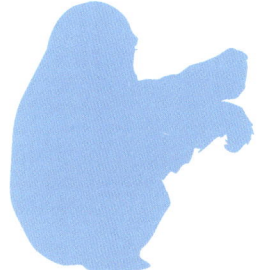

Princess was a female **ORANGUTAN** who lived in a wildlife reserve in Borneo and learned some sign language. One of her favourite pastimes was washing clothes – a skill she learned by copying people. She liked to lather up lots of soap and rub it into the cloth before eating the bubbles.

NIM CHIMPSKY, born in 1973, was selected to take part in a project to learn more about how chimps communicate. The baby chimp was given to a human family, who raised him as if he were a human child. They even dressed him in toddlers' clothes. Unsurprisingly, Nim behaved more like a chimp than a human, and after a few years the family could no longer cope with a powerful male ape. After being moved to a number of different primate research centres, eventually Nim was moved to an animal sanctuary in Texas, USA.

ORPHANED WILD APES can need help for a variety of reasons. They are sometimes sold as pets, but when they get too big they are abandoned. They may have been displaced by deforestation, or injured at the hands of humans. In order to return to a natural life in the world, the humans who care for them have to look after them like a mother, but also teach them how to be apes. This process is called 'rehabilitation' and it's a complex process.

The name 'ORANGUTAN' comes from the Malay words for 'person of the forest'.

BABY GORILLAS, for instance, are paired with a human care-giver who spends the day with them in the forest, foraging for food. They even sleep together, just like a mother and infant might. As soon as possible, the babies are handed over to a new gorilla family, so that experienced gorilla mothers can adopt them. They will gradually be given more independence, starting to spend more time in the forest and away from human contact until they become self-sufficient, and are able to live a free life.

BOB
— THE —
STREET CAT

When James Bowen found a hungry, injured cat, he couldn't walk away from an animal in need, even though he was struggling to feed himself. The ginger cat's green eyes glowed brightly in the darkness, and James's heart melted.

At the time, James was living in sheltered housing in London, and being supported by a charity. He was trying to pull his fractured life together after a very difficult start. As a child, he'd coped with his parents' divorce and tough times when his family kept moving home. That made friendships difficult, and led to endless bullying at school. Unhappy and restless, James had dropped out of education and for years he struggled with drug abuse, homelessness and mental health problems.

When James first met Bob, the ginger cat who was to change his life, he was working as a busker in London's Covent Garden. With support, he had cleaned up his life and stayed away from drugs. Playing the guitar on the streets helped James to earn just enough money to scrape by. But the young man had lost touch with his family, and endured long spells of loneliness and depression. Bob found James at just the right time for both of them.

Bob was skinny, sickly and suffering from a large wound to his leg. Despite the cost, James took Bob to a vet for treatment and looked after him while his infections healed. After a few weeks, Bob was growing fit and healthy, so James prepared himself to say goodbye to his new friend. He didn't know where Bob had come from, but if he was a street cat the chances were strong that Bob would want to return to his life outdoors.

But James was wrong and Bob turned out to be the best friend he could have hoped for, repaying his love and care a thousand times over. He refused to leave James's side, even following him onto the bus one day when James was heading to work.

Bob quickly became a popular addition to James's performances. He sat on James's shoulder as he walked through London's streets, and sat in his guitar case while James sang. He was a hit with tourists and Londoners – and James saw his earnings more than double when Bob was with him. More importantly, James knew that Bob needed him as much as he needed Bob, and their strong bond helped him to make new friends, and to manage his mental health difficulties.

Living in a city famous for its film-makers and storytellers, perhaps it's not surprising that Bob and James soon became a hit with the media as well as with the public. James was invited to share the story of his life with Bob in a book called *A Street Cat Named Bob*, and a film followed in 2016. Bob even played himself in some of the scenes! In his book, James wrote, 'Bob is my best mate and the one who has guided me towards a different – and better – way of life … Everyone deserves a friend like Bob.'

More bestselling books followed, and James used his success to support charities for homeless people and animals – especially cats, of course! Bob sadly died in 2020, but James will never forget what he owes his loyal feline friend.

HEROIC CATS

CRIMEAN TOM was a cat who lived with British soldiers during the Crimean War (1853–56). The soldiers were running dangerously short of food, but noticed that Tom always seemed well fed. One day they followed Tom, who led them to a hidden store of food he had found, and helped them to avoid starvation.

SMUDGE, a beautiful green-eyed tabby, certainly proved his worth as a guard cat. When bullies were attacking the two little boys he lived with, Smudge leapt out of the bushes, hissing, and jumped on the bullies, who ran away.

Some people believe their cats have alerted them to illnesses they didn't know they had. When Angela Tinning's pet cat **MISSY** began to claw at her chest, she thought the behaviour was so odd it was worth checking out. Her doctor found she had the early stages of cancer, and Angela thinks that Missy may have saved her life.

When **BABY THE CAT** smelled fire in the family home he didn't run for the door. Instead, he tried to find his owners to alert them to the danger. Josh and Letitia were fast asleep until Baby jumped on them!

SIMON was a much-loved cat who lived aboard a British boat, the HMS *Amethyst*, in the late 1940s. When the boat was attacked, Simon was badly injured. Thankfully, he survived his wounds and went on to save the sailors' store of food when it was overrun with rats. Simon's rat-killing efforts were awarded with a medal.

CATS
DID YOU KNOW?

WHAT IS A CAT?

Cats belong to a group of animals called felids or felines. They have strong, flexible bodies and small, round faces with short muzzles. Cats have superb senses and are able to see well in dim light. They use their sharp claws and teeth for catching and killing prey.

DO CATS HAVE NINE LIVES?

Cats have just one life, but they are so good at getting themselves out of scrapes that they have earned their reputation as super-survivors. One of their most useful skills is the ability to 'right' their body. They twist their spines when they fall, so they can land safely on four paws. Cats begin to develop this skill when they are just three weeks old.

HOW DO CATS FIND THEIR WAY HOME?

All cats are good at learning the area around their own homes, using visual clues as well as their sense of smell to mark and recognize their territory. They also have good memories. Some pet felines, however, have demonstrated astonishing powers that have left scientists mystified. Holly the cat, for example, went missing when her owners took her on holiday with them, travelling over 300 kilometres from their home. She turned up two months later having found her way back alone. No one knows how she did it.

DID CURIOSITY KILL THE CAT?

Cats are famous for their nosiness. There is nothing they like more than exploring high places, dark corners and tight spots. They use their sensitive whiskers to assess the width of a hole they plan to investigate, and rarely squeeze their bodies into places they can't get out of.

WHY DO CATS WASH SO MUCH?

Cats can spend half of their waking hours washing themselves. And they don't just groom themselves, they will happily lick clean other pets – and humans – if they get the chance. Licking helps a furry animal to find and clean any injuries, and remove any fleas or other bugs. It also helps a cat to keep cool on a hot day, and it feels good!

WHY DO CATS PURR?

Cats purr when they are content, but they have other good reasons for making this strange sound. It releases endorphins (special brain chemicals that help to reduce pain), so a purring cat may be injured or unwell. Some scientists believe that the vibrations of a purr may encourage healing.

HOW LONG DO CATS SPEND ASLEEP?

All members of the cat family enjoy plenty of naps, and pet cats are no exception. Newborn kittens spend almost every hour asleep, but even when they are adults they will enjoy snoozing for up to 18 hours a day.

WHY DO CATS HAVE SUCH ROUGH TONGUES?

Cats' tongues are rough and covered with tiny backwards-facing spines that contain grooves, like little scoops. This makes them ideal for scraping meat off bones – but it also helps cats to drink lots of water at a time and spread saliva over all their fur. That means a cat's tongue can work like a sponge and a scrubbing brush at the same time.

OR7
— THE —
LONE WOLF

When a young grey wolf set out on an epic journey, he had no idea how far his adventure would take him – or that it would earn him a place in the history books. All he knew was that it was time to leave his pack and find his own way in life.

It was September 2011 and the two-year-old wolf lived with his pack in the wilds of Oregon, in the north-west region of the United States. Wolves had been heavily hunted for centuries. They were so close to being wiped out in most of North America that biologists were studying this pack in the hope of protecting the remaining wolves. They fitted all the cubs with radio transmitter collars so they could follow their movements from afar. OR7 was so named because he was the seventh Oregon cub to be collared.

Driven by the need to find his own mate, and deep instincts that are far beyond human understanding, OR7 trekked onwards through the winter. He crossed ancient lava flows, climbed mountains and explored woodlands as he journeyed over 1,900 kilometres until, in late December, he reached California and became the first wolf in the state since 1924. Other wolves followed and OR7 eventually found a mate, and started a new pack.

OR7's journey captured the public's imagination, and he did more to help save his species from extinction than any other wolf. Celebrated by conservationists and wildlife lovers, wolves are now welcomed into some of their native homelands.

JACKALS

Jackals hunt at night, and on cool evenings they can run for several hours at a time, on the trail of prey. They look for small mammals and birds, which they pounce on. Jackals sometimes hunt in packs. They are found in Southern Europe, the Middle East, Africa and parts of Asia.

COYOTES

Most coyotes live alone, but sometimes they form small hunting packs. They live in Central and North America, and can cope with the heat of a tropical forest in Costa Rica or the freezing winters of Alaska. Coyotes pull fish out of rivers and climb trees in search of food, but they also eat carrion (dead animals) that they find or steal from other predators.

THE WILD DOG FAMILY

All wild dogs belong to a large family of animals called CANIDS. They have slender bodies, long legs, feet with four toes and bushy tails. Canids are clever hunters. They adapt quickly to new situations and often live in groups called packs.

WOLVES

Did you know domestic, or pet, dogs are canids and are all descended from WOLVES?

Wolves are the largest canids, and the grey wolf is the largest of them all. These beautiful carnivores were once common, but now they are found mainly in Russia, North America and China. Most types are grey with flecks of black, but there are white, brown, black and sand-coloured wolves as well. They work together to hunt animals that are larger than themselves.

AFRICAN WILD DOGS

African wild dogs live in large family groups and they share the task of caring for the cubs. They also work together to hunt, and this skill allows them to target big animals, such as wildebeest and zebras. These clever dogs live in Africa but they are endangered. There are approximately 1,400 of them left in the wild.

DINGOS are wild dogs that live in Australia. They are probably descended from domestic dogs. They often hunt in packs and are often regarded as pests that attack farm animals and spread diseases. Dingos also hunt small kangaroos, rabbits and birds.

FOXES

There are 23 species of fox and they are found throughout most of the world. The red fox has probably the largest natural reach of any land mammal, apart from humans. Foxes owe much of their success to their broad diet; they will eat almost anything they find.

FENNEC FOXES are the smallest wild canids. They live in the desert and have huge ears that they use to listen out for the bugs that they hunt. The soft fur on their feet stops them burning on the hot sand.

The **DIRE WOLF** was the size of a modern grey wolf but it had much bigger teeth. It died out about 10,000 years ago.

WILD DOG STATS
Size: 24–150 cm body length
Habitat: Woods, forests, grasslands, deserts
Found: Worldwide
Number of species: 36

COMING HOME

Many animals are being helped to return to their native lands, including moose, European bison, beavers, brown bears, griffon vultures, golden jackals and Iberian lynxes – one of the world's most endangered cat species.

ARABIAN ORYX are a stunning sight in the desert. They have white fur and wide hooves that help them to walk on shifting sand, and two long, rapier-like horns on their head. The horns – which can grow to more than 60 centimetres long – are so spectacular that people hunted these gentle antelope so they could own them. By 1972 the last wild oryx had been shot dead. Captive Arabian oryx have been bred so that herds could be returned to the wild, and there are now up to 1,200 living free in the Arabian Peninsula.

GREY WOLVES have been returned to the famous Yellowstone National Park in the United States. Now there are approximately ten packs of wolves in the Yellowstone area.

Not all stories are a great success. **RED WOLVES** had gone extinct in the wild by 1980. In 1987, a small population of captive red wolves was reintroduced into eastern North Carolina, where they compete with coyotes for prey. Fewer than 20 red wolves now live in the wild and it's an uphill struggle for scientists to guarantee their survival.

There is only one type of truly **WILD HORSE** left alive in the world: **PRZEWALSKI'S HORSE**. It was named after a Russian explorer who first described it grazing in Mongolian and Chinese grasslands in the late 19th century. It was extinct in the wild by the 1960s, but horses were being bred in captivity. Recently, captive herds have been reintroduced to the wild, and thousands of mature horses now live free.

LEADERS OF THE PACK

Grey wolves were once the most widely distributed mammals in the northern hemisphere. They are clever and agile predators, and much of their success relies on their ability to live and work as a team – the wolf pack.

The Alpha Couple

Each pack of wolves is led by two wolves: the alpha pair. They are bonded for life, and they are the only wolves in the pack that have cubs. The other wolves serve the alpha pair, and even help the alpha female to raise her young.

Cub Love

An alpha female can give birth to up to 14 cubs at a time and for the first few weeks the cubs, who are born blind and deaf, stay in the den that she has prepared for them. Soon the cubs are able to eat solid food that is provided for them by other members of the pack, who even chew it for them first!

Hunting Pack

Co-operation is essential to a wolf pack's hunting success. It can pursue caribou for days at a time, looking for an individual that is smaller or weaker than the rest of the herd. By attacking as a team a pack of wolves can bring down large animals – enough to feed the whole family. Younger wolves join in the hunt, watching and learning.

Dog Talk

The ability of domestic dogs to follow commands and interpret their owners' moods is something they have inherited from their wolf ancestors. To live as a family, a wolf pack relies on complex communication between the members. Body language, such as crouching or rolling over to expose their soft bellies, is an essential part of their 'language'. Wolves howl to tell other members of the pack where they are, and to warn rivals from neighbouring packs to keep their distance.

KEIKO
— THE UNLUCKY —
ORCA

The name Keiko means 'lucky one' in Japanese, but Keiko was never very lucky. He began his life in the chilly, fish-filled waters around Iceland where he was part of a family, or pod, of orcas. Like other toothed whales, orcas are sociable animals and young males sometimes stay with their mothers for all their lives. That wasn't to be Keiko's destiny.

Captured and put in an aquarium, Keiko began life in captivity when he was still a youngster. By 1985 he was living in a Mexican aquarium, where he performed for the public. His tank was too small, the water too warm and the lifestyle unsuitable for such a clever animal.

Keiko's life changed when he was used in a film to portray an orca that lived in an amusement park and escaped back to the ocean. *Free Willy* was very successful, but the film's fans were sad to learn that Keiko, its star, was still in captivity, so they raised money to improve his life. A bigger tank was built for Keiko, and scientists prepared him to be returned to the wild.

Keiko was flown back to Iceland, and by 2002 he was free in the ocean of his birth. Sadly, Keiko had come to rely on humans for both food and company, and he was never able to join a wild pod. He died soon afterwards, but the story of his sad life has challenged people to think about the lives of animals, especially marine mammals, in captivity.

WILD WHALES
DID YOU KNOW?

Whales and dolphins belong to a group of aquatic mammals called cetaceans. This family includes some of the most intelligent animals on the planet, as well as the biggest animals that have ever lived – blue whales.

HOW MANY TYPES OF WHALE ARE THERE?

There are 76 species of toothed whales, including dolphins and porpoises. The remaining 15 species of whale are baleen whales. They eat small marine animals by straining seawater through sieve-like 'baleen' plates in their mouths.

ARE WHALES CLEVER?

Scientists are only just beginning to understand how intelligent whales are. They can find food by sending out clicks of sound that bounce off objects nearby and then analyze the echoes that bounce back to them. They communicate with a range of sounds. Dolphins even use particular whistles for each other – just like using a name.

ARE DOLPHINS STILL KEPT IN CAPTIVITY?

Although many facilities – such as amusement parks – have closed down in recent years, there are still thousands of dolphins kept captive around the world, including young dolphins that are taken from their wild families.

In 1978, a group of **ORCAS** took on an enormous challenge when they decided to attack a young blue whale. Blue whales can grow to 30 metres long and even youngsters are huge. By working together, the orcas relentlessly pestered the whale, biting it repeatedly. It escaped, but probably died of its injuries soon afterwards.

ARE KILLER WHALES DANGEROUS?

Killer whales are also known as blackfish and orcas. They pose no threat to humans in the wild, but they are deadly predators of fish, seals, sharks and even other whales. Pods of orcas work together to create waves that knock seals off ice floes … and straight into the mouth of a waiting orca.

WHY ARE ORCAS CALLED KILLER WHALES?

Orcas are superb hunters that teach their young how to kill. A mother orca shows her calf how to wallop a group of fish with its tail to stun them, and demonstrates how to swim under a seal and flip it up into the air so it lands in her mouth. She will even take her young on hunting trips and show it prey before moving out of the way and letting the youngster practise catching and killing.

Many species of **CETACEANS** are endangered and in 1986 most countries in the world agreed to stop hunting whales and dolphins. Despite this, several countries have continued to catch and kill cetaceans.

A group of cetaceans is called a **POD**. An orca pod can number up to 50 and it's a close-knit family.

SUPER CETACEANS

TEAMWORK

In the seas off Laguna, Brazil, humans and wild bottlenose dolphins work together to catch fish, and they've been co-operating this way for at least 100 years. Fishermen stand waist-deep in water or in small boats and wait for the dolphins to chase shoals of mullet towards them. The water is too murky for the fishermen to see the fish, so they wait until the dolphins signal to them – by slapping the water with their tails – before casting their nets. The dolphins feast on fish that aren't trapped by the nets, so everyone in the team benefits.

TILIKUM THE ORCA

Tilikum was an orca that became famous after he was involved in three tragic events at aquariums where he was kept in captivity. Tilikum was taken from the wild when he was just two years old and trained to perform for the amusement of tourists. Two trainers died after accidents involving Tilikum and a tourist who had sneaked into Tilikum's pool one night was killed. Wildlife experts have argued that intelligent animals, such as orcas, are frustrated by an unnatural life in captivity.

SAVIOUR DOLPHINS

In 2004, a group of swimmers in New Zealand were saved from the jaws of death – by a pod of bottlenose dolphins. Rob Howes had taken his teenage daughter and two of her friends for a lifeguard training session in the ocean when they suddenly became aware of a pod of dolphins circling them. When Rob tried to swim away the dolphins herded him back into the centre of the group – and that's when he noticed a great white shark nearby. The dolphins continued to protect the swimmers until the shark gave up and swam away.

A DIVER IN PERIL

When Yang Yun felt a beluga whale grip her leg in its jaws she felt a moment of terror, before realizing that the whale was saving her life. Yang was taking part in a free-diving contest in an aquarium. The deep tank was home to Mila, a beluga whale, and filled with water that was chilled to mimic the whale's natural home in the Arctic Ocean. Yang was diving without breathing equipment, but when she suffered cramp she couldn't swim back to the surface and feared she would drown. That's when Mila got involved. She gently held Yang's leg in her mouth as she forced the young woman up to the top of the pool.

GREAT WHITE ATTACK

In 2007, Todd Endris was surfing in the Pacific Ocean when an enormous great white shark mistook him for a seal, and launched an attack on the young man. Within minutes of being bitten, Todd was surrounded by a group of bottlenose dolphins who kept the shark at bay while his friends saved him. He needed more than 500 stitches, but survived. Todd was surfing again just six weeks after the attack. He later argued for the need to protect great white sharks; he pointed out that he had foolishly chosen to surf in a marine sanctuary where sharks live.

BRAVE BALTO
— BEATS THE —
BLIZZARDS

In January 1925, Balto the sledge dog was not considered a hero. He wasn't even considered a particularly impressive or strong animal. Yet his courage and stamina helped to save many children's lives.

Balto lived in Alaska, where winters are fierce and long, and dog-pulled sledges are an efficient way to transport people and goods across the snow-caked landscape. Doctors in the small town of Nome had discovered an outbreak of diphtheria – a deadly disease that can make children ill very quickly. There was no time to lose. The doctors urgently needed anti-diphtheria serum, but the nearest supplies were far away in Anchorage.

The serum was taken by train to the town of Nenana, but from there the only way to reach Nome – a distance of about 1,000 kilometres – was by sledge. Teams of mushers (sledge drivers) and hardy dogs raced in relay across the frozen land where the temperature was as low as -40°C. On 1st February, it was the turn of Balto's team of dogs to take over, led by their musher Gunnar Kaasen.

The weather worsened, and the team battled snow blizzards through the night. The winds were strong enough to knock the dogs off their feet but they persevered, following Balto as he led the way. It took the dogs almost ten hours to cover 85 kilometres, but finally they arrived in Nome. It had been a life-or-death race, but Balto and his canine companions had overcome nature's most extreme weather to save a town's children.

SNOW DOGS

The first breeds of dogs put to work by humans were wolf-like and used to life outdoors, even in long, cold winters. They were bred to herd livestock and protect farmers' animals from wild beasts. Today, there are many breeds of dog that still thrive in cold places, and many of them do important work.

AVALANCHE!

An **AVALANCHE** is a sudden movement of a large amount of snow and ice down a mountainside. It can occur without warning, and the snow can engulf roads, homes and people.

Dogs have been trained to help find people buried in avalanches. Working with a human handler, an 'avy dog' has to scour the snow, sniffing for human smells. Just one dog can cover an area eight times faster than a team of 20 humans. Speed really matters as a skier buried in an avalanche is likely to suffocate or freeze to death in less than an hour. When an avy dog smells a human it starts to dig, and then humans join in with shovels.

One of the most famous snow-rescue dogs of all time was Barry the St Bernard. In the early 19th century, he is believed to have saved more than 40 people in the Alps, including a small boy whom Barry discovered asleep in a cavern of ice. Barry carried the child to safety on his back.

OTHER SNOW ANIMALS

Siberian Huskies

Siberian huskies closely resemble their ancestors, grey wolves. They have been long used as sledge dogs owing to their energy, fortitude and ability to cope with severe weather. It takes great skill and experience to train a husky, but they are loyal workers and can stay outside in frosty or snowy conditions for many hours.

Bactrian Camels

Wild two-humped Bactrian camels live on the vast plains of Central Asia. They can survive long periods without water and they grow thick, shaggy coats for the winter, but lose them in messy clumps when spring arrives. Bactrian camels were first domesticated more than 4,000 years ago and are used to carry loads.

Yaks

Yaks can grow to an enormous size – up to 2 metres tall at the shoulder and weighing more than 800 kilograms. They can keep warm in their mountain homes thanks to thick undercoats of soft brown fur, and top coats of long black hair. Yaks live in China, Nepal, Mongolia and parts of Central Asia. They are members of the cattle family and for centuries they have been bred by people who live at high altitudes. They are used as working animals, and for their meat and milk. Yak dung is burned as a fuel instead of wood on the plains of Tibet, where no trees grow.

Llamas

Llamas are sturdy, adaptable camelids that manage to survive in the snow-topped Andes, in South America. Their fur is used as wool for blankets and clothes, and they are strong enough to carry loads. Llamas like to live in groups, or herds, and their sociable behaviour makes them ideal for guarding flocks of sheep. They alert shepherds to any predators lurking nearby and will even kick or spit at anything that frightens or threatens them!

ASIAN ELEPHANTS have long worked as pack animals, but African elephants are almost impossible to train. Asian elephants have been used to plough fields, knock down trees to clear land for farming and to carry people or building materials. Nowadays, wildlife experts discourage the use of these majestic, intelligent and rare animals as beasts of burden.

DROMEDARY CAMELS were used as working animals during the First World War. In India, they were fitted with special baskets so wounded soldiers could be loaded onto their backs. The camels then carried the injured men to safety. In the Middle East, soldiers from New Zealand rode camels to reach the battlefront.

WHAT IS A BEAST OF BURDEN?

Throughout history, humans have used some animals, sometimes called 'beasts of burden' or 'pack animals', to do the hard work for them, from pulling carts, sledges and ploughs to carrying heavy loads on their backs.

In the far north, around the Arctic, **REINDEER** (also known as caribou) have been used as pack animals for thousands of years. These large, strong deer can tolerate the extreme cold, so they are used for pulling sledges in the winter and carrying loads on their backs in the summer.

ASSES or **DONKEYS** are probably the oldest-known pack animals. People have used these sturdy and strong members of the horse family to carry things for more than 5,000 years.

DOCTOR DOGS TO THE RESCUE!

Diagnosing diseases and curing illnesses isn't just down to the scientists and doctors. Animals can play an important part too!

Dogs have such a superb **SENSE OF SMELL** that they can even sniff signs of disease in humans. Some breeds are so sensitive to smell they can detect odours at concentrations of one or two parts per trillion. That's roughly equivalent to one teaspoon of salt in two Olympic-sized swimming pools.

Medical-alert assistance dogs are able to detect the **WARNING SIGNS** in some medical conditions, such as diabetes and epilepsy. They can smell small changes in a human's scent, which alert them to life-threatening events. The dogs can then warn their owners to seek treatment.

Malaria is a deadly disease that kills hundreds of thousands of people every year, especially children. However Lexie, Sally and Freya are dogs who may be able to change the future for Africa's most vulnerable people. They have been **TRAINED TO DETECT** if children are infected with the early stages of malaria just by sniffing their socks! If the disease is detected early enough, it can be treated before it harms the child.

Steven was diagnosed with type 1 diabetes when he was three, and like many children with diabetes it proved difficult to keep his blood-sugar levels steady, even though his parents were testing him more than ten times a day. Steven's **MEDICAL ALERT** dog, Molly, has changed his life. She stays by his side and lets Steven or his parents know if his blood-sugar levels need checking.

Scientists are currently working with medical **DETECTION DOGS** to discover if they can be reliably used to detect the early stages of cancer.

DIGIT THE GORILLA — IN THE — MIST

When Dian Fossey was growing up she struggled to make friends, but she was always happy in the company of animals. She was passionate about them, but her great love was to be the cause of her untimely and violent death. Thanks to a gorilla named Digit, however, her life and work were not in vain.

Dian had trained as an occupational therapist but she was restless for change. When she was offered the opportunity to leave her home in the United States and move to Africa to work with gorillas, she leapt at the chance. The plan was for Dian to live among rare mountain gorillas and study them.

It was 1967 when she arrived in the cloud-covered Virunga Mountains of Rwanda. With the help of local townspeople, she established a camp high on the slopes of the mountains that were cloaked in rainforest and mist. During the day, Dian and her Rwandan team trekked through dense undergrowth, often in pouring rain, to track down the shy gorillas. Once she found them, she would sit nearby, watching them and making notes.

Dian's life was far from idyllic. Living conditions were hard; it could rain for days without end and Dian was often unwell. But what made her most unhappy were the poachers. Zoos were offering money for baby gorillas, and tourists liked to buy souvenirs of gorilla hands to take home. As a result, poachers were tracking

the gorillas too, and killing or capturing them. The gorilla population was heading towards a catastrophe and Dian predicted that the whole species could be wiped out in just a few years.

Dian was entranced by one young gorilla in particular. She had named the young male Digit, and described him as a 'playful little ball of fluff'. Over the next ten years they spent many hours together. Digit and his family were content to let Dian sit amongst them as they chewed leaves and played, and Dian grew to love Digit more than she had ever loved a fellow human.

In 1977, poachers targeted Digit's family with tragic results. As a young silverback, Digit's job was to protect his whole family. He did his duty, and he fought back against dogs and spears to give his family, including his pregnant mate Simba, time to escape. But Digit's heroic actions cost him his life.

After a time spent in terrible grief, Dian and her colleague Ian Redmond, who had discovered Digit's body, decided it was important to carry on their research work, and gather more support for their goal of saving mountain gorillas. At the time, Dian estimated there were just 200 mountain gorillas left, and she knew that Digit's shocking death could be used to help publicize the dangers they faced.

Dian and Ian spread the news of Digit's death around the world, and money was raised to help set up conservation projects. However, not everyone was happy with Dian and how she planned to protect the gorillas. The poachers wanted her out of the mountains, and on 26th December, 1985, Dian Fossey was murdered in her cabin. Her killers were never captured, but the work she had started continues today.

There are now more than a thousand mountain gorillas and they are the only type of great apes that have increased in numbers in recent years. When Digit died he didn't just save his family; he may have saved his species.

HUMAN APE HEROES

Louis Leakey (1903–1972) was fascinated by the early history of humans and believed that understanding how modern primates live would help him learn how humans evolved. He set up three camps and chose three young women to lead the research: Dian Fossey went to Rwanda to study mountain gorillas, Biruté Galdikas went to Borneo to study orangutans and Jane Goodall went to Tanzania to learn more about chimpanzees, our closest cousins.

Biruté Galdikas

When Biruté Galdikas arrived in Tanjung Puting Reserve in Borneo in 1971, she was entering one of the last great wildernesses on Earth. There were no roads, no electricity and no telephones.

The scientist was told by wildlife experts that her dream of studying orangutans in the wild 'couldn't be done' because the orange apes were hidden in dense forests with deep swamps. She proved them wrong, and Dr Galdikas worked for more than 40 years, conducting the longest continuous study by one researcher of any wild mammal in the world.

Dr Galdikas has worked tirelessly to understand how orangutans live, and to ensure they have a future. She set up a charity, Orangutan Foundation International, in 1986 to preserve their habitat and raise money for their conservation.

Jane Goodall

When Jane Goodall arrived in Tanzania in 1960 to study wild chimpanzees, she had little more than a notebook and a pair of binoculars. But her dedication, courage and perseverance led to a lifetime of ground-breaking research and conservation work.

As a child, Jane had dreamed of working and living among wild animals in Africa. Her dream finally came true when Dr Leakey employed her to study the chimps at Gombe Stream Chimpanzee Reserve in Tanzania. She was the first researcher to observe chimps eating meat and using tools, and worked ceaselessly to care for orphaned chimps.

Dr Goodall has set up organizations to support the communities who live alongside chimp habitats. She travels the world to spread the word about the threats that chimps face and preserves their natural environment. She has also endeavoured to improve the lives of chimps held in zoos around the world.

GORILLA DANGER

Large male gorillas, called silverbacks, are incredibly strong and fearsome. They are equipped with huge fangs that could inflict a deadly bite on a human, but both male and female gorillas can be very gentle, too.

Jambo

In 1986, a five-year-old boy called Levan Merritt fell into the gorilla enclosure at Jersey Zoo. As he hit the ground, Levan broke his arm and fractured his skull. As the little boy lay motionless, Jambo, a silverback, came over to investigate, and horrified onlookers watched. Jambo was intrigued by the injured child, and gently stroked him on the back. He even placed himself between Levan and other curious gorillas until zookeepers managed to rescue the boy.

Binti Jua

In 1996, a three-year-old boy fell into a gorilla enclosure at Brookfield Zoo, near Chicago, USA. He lay unconscious on the floor, but a young female gorilla called Binti Jua came to help him. She held the boy gently in her arms and carried him away from other animals so that humans were then able to rescue him.

Harambe

In 2016, a three-year-old boy climbed into a gorilla enclosure at Cincinnati Zoo, USA. A large male gorilla called Harambe dragged the boy and pushed him around. The gorilla was further upset by the screams of onlookers. Fearing for the boy's safety, zoo officials made the difficult decision to shoot Harambe. Wildlife experts agreed that it was the right decision, as the boy's life was in danger.

SILVERBACKS need to be powerful to protect their families, and when they are threatened they put on a great display of strength, standing tall and beating their chests while hooting and grunting.

WILD GORILLAS

Gorillas are great apes that live in the forests of Central Africa. They are our closest living relatives after chimpanzees and bonobos, and gorillas are much-loved animals, perhaps because they remind us of ourselves.

A group of gorillas is called a TROOP and it usually contains a number of families. There is one big male who is the leader. As other males mature they leave the troop to start their own families.

MOUNTAIN GORILLAS have longer, thicker fur than other types of gorilla because their forest home is located on mountainsides that are shrouded in damp mists.

BABY GORILLAS use their hands and feet to cling to their mothers as they move around. They feed on their mother's milk until they are about three years old and they learn which plants are safe to eat by watching older gorillas find food.

Gorillas TALK to each other using deep noises that sound like burps!

Gorillas are PEACEFUL ANIMALS that spend most of their time lazing on the ground, playing, sleeping and eating plants. The biggest threat to them – apart from humans – are leopards that prey on young gorillas, and silverbacks from neighbouring troops.

Today, Digit's family and other mountain gorillas are protected by the communities of people who live nearby. Tourists are invited to visit the gorillas in their NATURAL HABITAT and the money raised by these projects is used by local people to fund their children's education, forest conservation and GORILLA-FRIENDLY farming.

GORILLA STATS

Size: Up to 180 cm long

Habitat: Forest

Diet: Fruit, leaves, seeds, bark, roots and some bugs

Lifespan: Up to 50 years

WOJTEK
— THE —
SOLDIER BEAR

The fighting had come to a halt and the field of battle fell briefly silent, shrouded in a fine mist. A soldier stopped in his tracks when he spied an extraordinary sight. He grabbed his sketchpad and pencil from his bag and began to draw. He was sketching a huge brown bear that was gently carrying an artillery shell in his strong arms.

The story of Wojtek the Syrian brown bear began two years earlier in 1942, during the Second World War. A boy had rescued a bear cub after its mother had been shot, and sold it to Polish soldiers. They named it Wojtek and he soon became a much-loved pet. Since bears were not allowed to travel with the troops, the Poles made him a 'Private' so they could issue Wojtek with the necessary papers.

In 1944, the soldiers found themselves in Italy, supplying food, ammunition and artillery shells to British and Polish troops. Wojtek had grown taller than most of the men – and he was stronger than any of them.

One day in battle, Wojtek decided to help. His human colleagues were desperately loading up trucks with supplies for the men who were fighting. Wojtek stepped forward, with arms outstretched. Following his cue, the soldiers loaded the bear up with vital supplies, and for days he carried heavy crates from truck to truck, unflinching as bombs exploded all around. Wojtek and the human soldiers he worked with had helped to win the battle, as well as the hearts of everyone who heard his story.

MYTHICAL, MAGICAL BEARS

The Ancient Greeks told a tale of a bear called **CALLISTO**. The bear had once been a beautiful nymph who bore a son, called Arcas. Arcas's father was Zeus, the king of the gods. When Zeus's wife discovered this, she was so jealous that she turned the nymph into a bear. Years later, Arcas was out hunting when he saw a great bear and decided to kill it. He didn't realize that the bear was actually his mother. He was about to release his arrow when Zeus saw what was happening and quickly transformed them both into stars in the northern sky to save them from their dreadful fate. Callisto became the stars that make up the Great Bear (also known as the Ursa Major constellation) and Arcas became the Little Bear (also known as the Ursa Minor constellation).

According to old Norse legends there was a race of super-humans, known as **BERSERKERS**, who could transform into bear-like warriors. It was said they had the strength of bears and fought with a vicious frenzy. They have given us the adjective 'berserk', which is used to describe a person or animal who is out of control with anger or excitement.

The Modoc people of North America tell a story about bears. When the chief of the sky spirits made the Earth, he made **GRIZZLY BEARS** too. One day, a grizzly bear found the daughter of the chief of the sky spirits and took her home. When she grew up, she married the son of the grizzly bear. Their children, who were half grizzly and half sky-spirit people, became the first Native Americans.

An Inuit story from Greenland reminds people that they need to respect the **POLAR BEARS** that live alongside them. It tells of an old hunter who killed a polar bear but prayed for it to come back to life. A little bear cub stepped out of the dead bear's body and the hunter took it home. He and his wife raised the cub as their son. When it grew up, the bear proved to be a good hunter and brought seals to the couple for them to eat. One day, the old man told his bear-son to kill a polar bear and bring back its meat for them to eat. The young bear did not want to kill his own kind, but obeyed his father, as that was his duty. When his parents sat down to eat the meat, their bear-son stepped outside and never returned. Without his help, the old couple could not hunt and no longer had any food at all.

The people of many Native American tribes believed bears had supernatural powers and could help cure illnesses. Groups of healers, or shamans, were called **'BEAR-DOCTORS'**. They watched bears to learn what foods would be good for people, and used the herbs, berries and leaves that bears ate in their medicines.

THE ARCTIC

is the region of the Earth that surrounds the North Pole. Its name comes from the Greek word 'arktos', which means 'bear'.

DANGEROUS BEARS
DID YOU KNOW?

ARE BEARS DANGEROUS?

Most bears are omnivores, which means they can eat a varied diet that includes both plants and animals. They rarely attack humans except to defend themselves or their young. Black bears are believed to be slightly more aggressive than brown bears, and are more likely to attack to protect their food.

WHAT ARE A BEAR'S LETHAL WEAPONS?

A bear's canine teeth can be 8 centimetres long, and its claws are long and knife-like, reaching up to 10 centimetres. A bear is also equipped with an impressive sense of smell and great strength. A single swipe with a mighty paw is enough to knock a grown man to the ground. Bears can also run surprisingly fast, so never try to escape a bear's attention by running away!

HOW DO YOU KNOW IF A BEAR IS GOING TO ATTACK?

Most bears are shy and would rather frighten people away than fight them. When threatened they warn people to stay away. These signs include growling and huffing and beating the ground with their paws. They stand up, to make themselves appear bigger, and bare their teeth as a final threat before making bluff charges. In this situation it is wise to calmly walk backwards, keeping an eye on the bear while retreating to safety.

WHY DO SOME BEARS ATTACK PEOPLE?

When a bear is threatened it can be very dangerous, especially a mother protecting her cubs. In some places, where humans have moved into bears' habitats, the bears have lost their fear of humans – and that makes them much more dangerous. Bears are intelligent animals, so it's not surprising they have learned that humans often carry food with them, or store food in tents, so they approach humans to scavenge food.

WHAT'S THE MOST DANGEROUS BEAR?

Polar bears are impressive predators and one of the few animals that hunt humans for food. Thankfully, they live in the cold Arctic, where they rarely encounter people. However, as the planet is beginning to warm up, polar bears are losing some of their icy habitat, and finding it harder to hunt. This forces them to travel further in search of food, and increases the chance of dangerous contact with humans.

WHICH ANIMALS ARE DEADLIER THAN BEARS?

Humans are far deadlier than bears. The number of people killed by bears in the whole of North America from 1900 to 2009 has been calculated to be 63, but humans kill at least 30,000 bears worldwide every year, and some species of bear are endangered.

OLD BLUE
ON THE BRINK OF
EXTINCTION

In 1980, Wildlife Officer Don Merton was given an almost impossible challenge: he was instructed to save an entire species from extinction. There were just five Chatham Island black robins alive, and failure stared Don in the face.

The story began years earlier, when people settled on islands near New Zealand and set to work, felling trees and sowing crops. They had brought cats and rats with them, and soon these predators were rampaging across the damaged habitat, killing birds that could neither defend themselves nor their chicks.

By the 1970s, only seven robins survived. Don refused to despair even when two more birds died, and he realized that only one remaining pair of birds could produce chicks – Old Blue and her mate Old Yellow. These robins usually live to the age of four and produce two chicks in a breeding season. At that rate, they were doomed.

Don had an idea. He took Old Blue's eggs and placed them in the nest of another native bird. The foster parents cared for the eggs, and Old Blue promptly laid two more. When her first clutch hatched, the chicks were returned to Old Blue so she and her mate could rear them, while foster parents incubated the second clutch of eggs.

Don's ingenious plan had worked. Old Blue carried on laying eggs that were fostered, and she lived to the extraordinary age of 14 years. Today there are nearly 300 Chatham Island black robins – all of them descended from Old Blue.

WHAT IS EXTINCTION?

Extinction means that all the individuals of one species of living thing have died. The rate of extinction is rapidly increasing because of human behaviour, such as the cutting down of forests and polluting oceans. However, some extinctions in the past have happened naturally.

Why Do Animals Become Extinct?

Animals face a lifelong battle for survival. Each species of animal must adapt and change to cope with a changing environment. Most extinctions are a natural part of the story of life on Earth. It is a process – called natural selection – that leads to new species of animals better suited to survive on an ever-changing planet.

Why Did Dinosaurs Die Out?

It is thought that about 66 million years ago an enormous meteorite smashed into Earth and brought about such massive destruction that most animals and plants died out. Those that did survive quickly adapted and thrived. If dinosaurs hadn't died out there wouldn't be such a huge range of birds and mammals alive today (and that includes humans!).

Past Extinctions

Most extinctions occurred long before humans evolved. We know about them from fossils. These are the remains of long-dead animals and plants that, over time, were turned to stone. The soft body parts of an animal are rarely fossilized, so scientists mostly study the fossils that have formed from hard body parts, such as bones, scales, teeth, claws and shells.

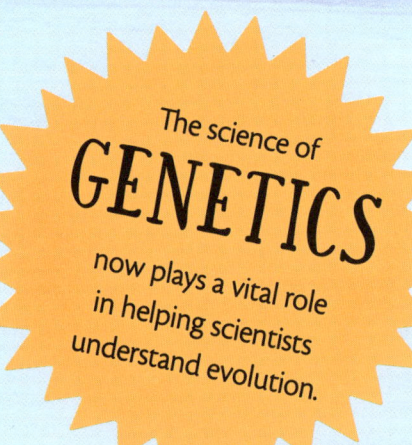

The science of GENETICS now plays a vital role in helping scientists understand evolution.

ALIEN INVADERS

Aliens are a major threat to wildlife – but not the ones from outer space. These aliens are all animals that have been introduced to habitats where they don't belong, often with catastrophic consequences for the native wildlife.

Cats and Rats

Rats have often travelled alongside humans by sneaking on board ships. Sailors used cats to catch rats. When ships landed on islands around Australia and New Zealand, these rats and cats escaped and caused great destruction. Large flightless parrots, called kakapos, were brought close to extinction after being hunted by cats and rats. Today there are about 155 of these birds left alive. They are protected on islands where alien species have been removed.

Asian Carp

These large silvery fish were brought to the United States in the 1970s to eat plankton in fish farms and keep them clean. However, some escaped after flooding and they quickly spread across the country and now compete with native fish for food and habitats. With no natural predators in their new home, they can quickly populate lakes and rivers.

Cane Toads

One of the most famous alien invaders is the American cane toad, which was introduced to Australia in the 1930s to combat pests that harmed crops. They produce a foul, toxic ooze from their skin which predators in their native habitat have some immunity to – but Australian predators are powerless against it. With nothing to halt their progress, populations of cane toads have exploded.

Brown Tree Snake

When brown tree snakes were accidentally introduced into Guam from Papua New Guinea in the 1950s, these small snakes caused the local extinction of more than half of the island's native bird and lizard species and two-thirds of the bat species. Without the birds and lizards to eat insects, insects have grown in numbers and have become pests that damage crops. The island's farmers now grow less food, and plants across the island have been affected because the birds, reptiles and bats that pollinated them are gone.

THE LAST OF THEIR KIND

Biologists believe that the Earth is now undergoing a mass extinction, and it's one that has been caused by humans. These animal superheroes are the last of their species, and unless we make some dramatic changes to how we care for wildlife, many more creatures will face extinction.

Lonesome George

When the body of Lonesome George was found on the morning of June 24th, 2012, the world mourned. George was the last Pinta Island tortoise, a species of giant tortoise that lived on the Galapágos Islands. Giant tortoises had long suffered at the hands of humans. They were hunted for food and their island habitats had been damaged by humans and farm animals. Before George was discovered in the wild in 1971, biologists believed that all Pinta Island tortoises had already become extinct. They tried hard to find George a mate, but it soon became evident he really was the last of his species. He was more than 100 years old when he died.

Martha the Passenger Pigeon

In 1813, an American pioneer called John James Audubon was travelling home when he saw an enormous flock of passenger pigeons appear on the horizon. He described the flock as being so great that it blocked out sunlight, and 'a buzz of their wings' continued for several hours as all the birds passed by.

At the time there were an estimated 3 billion passenger pigeons alive in North America. A century later, in 1914, the last one, named Martha, died alone in Cincinnati Zoo. Her species had been wiped out by hunting, as passenger pigeons were a popular food for humans.

The Last Northern White Rhinos

Najin and her daughter Fatu are northern white rhinos that live in captivity in Kenya. They are the last two of their type since the last surviving male, Sudan, died in 2018. The last wild population of northern white rhinos had already been hunted to extinction by 2008.

The Disappearing Toad

Golden toads lived in the humid cloud forests of Costa Rica. The last one was seen in 1989, and this species was finally declared extinct in 2008 when all hope had been lost of finding any more of these precious amphibians. They died out as a result of a fungal disease that was probably caused by climate change and pollution.

The Final Hawaiian Honeyeater

Jim Jacobi is thought to be one of the last people ever to have heard the Kaua'i' ō'ō sing, back in 1984. Jim recorded the song, and thinks the bird was drawn to the sound when he played it back, thinking it was another Kaua'i' ō'ō. Perhaps this bird was the last of its species, now declared extinct.

Goodbye, Baiji

The Yangzte river dolphin, or baiji, is the first species of dolphin that has been driven to extinction by humans. There were only 400 baijis left by 1980, but their numbers continued to fall rapidly and the last confirmed sighting was in 2002. Their extinction was caused by pollution and fishing that killed or trapped the dolphins as well as fish.

LIN WANG
— GOES —
TO WAR

When Lin Wang died in Taipei Zoo at the great age of 86, he was not just a much-loved Asian elephant: he was a hero who had survived great hardship during war.

Lin Wang's story began during the Second World War, when the Japanese Army used him to carry supplies and pull heavy loads in Burma (now known as Myanmar). In 1943, Lin Wang was one of 13 elephants captured by Chinese forces. He was obliged to serve new masters and continued his gruelling work. In 1945, the Chinese soldiers took the team of elephants on a trek back to Guangdong, a province of China. The trip was long and arduous, and six of the elephants died along the way.

Lin Wang was rewarded for his efforts with a long spell in a park, before he was put back to work. The brave elephant was sent to Taiwan where he pulled logs at an army base. Despite the difficult work, Lin Wang followed orders and did his duty, but by 1951 all 12 of the other elephants that had been captured by the Chinese forces had died.

It was decided that Lin Wang deserved to retire and in 1954 he was sent to Taipei Zoo in Taiwan, where he spent the rest of his life. He was popular with visitors who called him 'Grandpa' and every year the city celebrated his birthday. When Lin Wang died in his sleep in 2003, he was the oldest elephant in captivity.

ELEPHANTS
DID YOU KNOW?

DO ELEPHANTS HAVE FAMILIES?

Elephants live in families called herds. The eldest female, who is called the 'matriarch', leads each herd which is made up of mothers, daughters, sisters and aunts. Everyone helps to care for the youngest elephants and to find food and water. There are up to 15 adults in a herd, and when a herd grows too big it splits, but the two groups often meet up. Male elephants live away from the females, either in groups or alone.

ARE ELEPHANTS CLEVER?

There is no doubt that elephants are very clever animals. Along with great apes and dolphins, elephants share the rare ability to recognize themselves in mirrors. They are able to learn, play and help members of their herd who are injured or hurt. They can make and use tools – for example, elephants strip slender branches from trees and use them to flick flies away.

WHY DO ELEPHANTS RUMBLE?

Although elephants can communicate with trumpeting sounds, they also make low rumbles. This rumbling can travel more than 3 kilometres through the ground, and other elephants 'hear' the sound through their feet. They also communicate with their trunks, greeting one another by putting their trunks in each other's mouths. They stroke their young to calm them or slap them if they misbehave.

DO ELEPHANTS REALLY HAVE LONG MEMORIES?

Elephants do have long memories. It's the female leader's job to remember where to find waterholes, often years after last using them. Elephants also visit the places where members of their family have died, and gently touch the ground on the actual spot where their loved one lay.

WHY DO ELEPHANTS HAVE BIG EARS?

Elephants use their big ears to hear, but they have other purposes too. Elephants flap their ears to show they are angry, and their huge surface area helps an elephant to keep cool. When an elephant is hot, its warm blood rushes to the ears, and, as it flaps its ears, the heat is radiated away from the body.

DO ELEPHANTS MIGRATE?

Like many other animals that live on the grasslands of Africa, elephants migrate with the seasons to make the most of the new plant growth that follows the rainy seasons. Elephants in Mali, however, live in one of the world's most challenging habitats – deserts, where temperatures soar and water is hard to find. During one year they can migrate more than 480 kilometres.

WHY ARE ELEPHANTS ENDANGERED?

For centuries, elephants have been killed for the ivory of their tusks. They have also been taken from the wild, especially in Asia, to work for humans or to be kept in captivity. Today, Asian elephants are particularly threatened by the loss of their habitat.

ELEPHANT STORIES

JUMBO was one of the world's most famous elephants, and his name is now used to describe anything enormous. Jumbo was captured in Africa, in 1861, after hunters killed his mother. He was transported to London Zoo, where he was hugely popular with visitors who were allowed to ride on his back. In 1882, Jumbo was sold to P.T. Barnum, an American showman who wanted the elephant to star in his celebrated circus. Just three years later, Jumbo died after being hit by a train, but Barnum continued to make money by charging people to view exhibits of Jumbo's body parts, which he had preserved.

In 2010, a wildlife charity called Born Free campaigned to save two baby elephants, **MAKWA** and **KENNEDY**, from life in a North Korean zoo. The youngsters were safely returned to the wilds of Zimbabwe to live their lives in freedom.

RUBY was an Asian elephant who was born in Thailand in 1973 and sent to Phoenix Zoo in Arizona, USA, when she was less than two years old. One day, Ruby's keepers noticed she was using a stick to scratch marks in the sand, so they gave her brushes and paint and she was soon daubing colours onto paper. Ruby's painting career made her famous, and her paintings were sold for thousands of dollars.

SHIRLEY was an Asian elephant who was captured in Sri Lanka in 1944 and worked hard as a circus elephant until 1995. She was finally allowed to retire to an animal park in Georgia, USA, where she lived well beyond the life expectancy for a captive elephant.

HANNIBAL was a military leader who battled the Ancient Romans. One of his most famous achievements was leading his army across the Alps. He took about 40 elephants with him, but by the time he arrived in Italy only a few were left alive. Hannibal's last-surviving elephant was **SURUS**, who had just one tusk and was said to be the bravest of them all.

In 326 BCE, **ALEXANDER THE GREAT** confronted **KING PORUS** on the banks of the Hydaspes River. He was hoping to conquer India, but King Porus used 200 elephants to charge Alexander's army and send his soldiers running. However, Alexander returned and attacked the elephants with a flurry of arrows. The elephants panicked and stampeded, killing many of Porus's men and Alexander was victorious.

In 2014, poachers killed a huge African elephant with a poisoned arrow. **SATAO** had enormous tusks that, at 2 metres long, almost touched the ground. They made him a target for poachers who could sell the tusks for large amounts of money. News of Satao's tragic death spread around the world and helped to publicize the work that is being done to put an end to the illegal trade in ivory.

LEFTY
THE PIT BULL TAKES A BULLET

Lefty sensed trouble. It was the middle of the night, and on any ordinary night Lefty and her human family would have expected to stay safely tucked up in bed until morning.

But as soon as she heard unexpected noises, Lefty's ears pricked up and her nose twitched ... she knew there were strangers in the house and she feared for everyone's safety. All dogs can be super-sensitive when new people enter their home, but most pit bulls make terrible guard dogs. They are so friendly that they are more likely to lick a stranger than growl!

Four armed intruders had broken into the house. A deep instinct warned Lefty that something was wrong and she was quickly on her feet, ready to act. Lefty's 'dad' was by now out of bed, but when one of the intruders aimed a shotgun at him the brave little dog flew forwards in a flash. As she jumped between the gunman and her 'dad', the intruder fired his weapon and the bullet hit Lefty in her shoulder, throwing her to the ground.

The burglars left, but they had robbed the family of their money and the children were devastated at the thought of losing their beloved dog too. Lefty needed urgent surgery to remove her leg, which was too badly injured to repair. News of Lefty's bravery was shared in the local community and on social media. Dog lovers from all over were quick to offer their help and Lefty was soon on the mend. She returned home to her loving, and very grateful, family.

DOGS
DID YOU KNOW?

WHAT IS A DOG?

A dog is a domestic animal that is closely related to wolves and other members of the canid family. All pet dogs are descended from grey wolves.

WHAT IS THE TALLEST DOG?

Great Danes are the tallest breed of dog. A Great Dane named Zeus was the tallest dog ever measured, at 111.8 centimetres tall. However, ultra-big dogs often have very short lifespans.

WHY DO DOGS LOOK SO DIFFERENT FROM WOLVES?

Over thousands of years humans have bred dogs to change the way they look and behave. Dogs were chosen for their characteristics, such as spotty coats and playful personalities. Over time, those 'types' have been further developed and are called breeds.

WHAT IS THE SMALLEST DOG?

Chihuahuas have been bred to be tiny, and the smallest ones are less than 20 centimetres long. The shortest Chihuahua is a female called Pearl. She's just 9.14 centimetres tall!

HOW MANY BREEDS ARE THERE?

There are at least 300 breeds of dog. When dogs from two different breeds are mated, their offspring is known as a 'cross-breed'. A Labradoodle is a cross-breed between a Labrador Retriever and a Poodle.

WHAT IS A TOY DOG?

Breeds of dog that are small are called toy dogs. Pekingese, Papillon and Maltese are all types of toy dog.

WHY ARE SOME DOGS HAIRLESS?

Hairless dogs have been bred because they don't need grooming and they don't make as much mess as hairy dogs. However, they get cold and sunburned so need plenty of special attention.

WHY DO DOGS BARK?

Grey wolves howl and yap to communicate, and dogs bark for the same reason. They use their barks to get attention, to give a warning or when they are excited. They also use other sounds, such as howling and whimpering, especially if they are bored, lonely or in pain.

WHICH ARE THE FASTEST DOGS IN THE WORLD?

Greyhounds are the fastest dogs. They were bred for hunting and can outrun most other animals, with top speeds of 70 kilometres an hour or more.

WHICH DOG BREEDS HAVE THE BEST SENSE OF SMELL?

Dogs with long snouts have a better sense of smell than short- or snub-nosed dogs. Breeds with an exceptional sense of smell are grouped together and called 'scent hounds'. They include Bloodhounds, Bassett Hounds and Beagles.

WHAT ARE THE HAIRIEST DOGS?

The Puli and the Komondor are two extremely hairy dogs. Their fur grows in long tassels or cords that reach the ground. They were bred to herd or guard sheep, and they are so furry they could easily be mistaken for sheep.

MEDALS FOR HEROES

When an animal goes the extra mile to save a life, surely it deserves a medal? Which of these superheroes would you give an award to?

CANARIES have saved the lives of many men working in coal mines. Toxic gases are sometimes present in mines but, because they are odourless and invisible, miners would sometimes die before being able to escape. The men discovered that if they took these small birds with them, they would quickly fall ill if there were any lethal gases in the air, giving the men time to return to the surface, and fresh air.

Scarlett the **BEAGLE** spent months in a research laboratory where she was used in animal experiments. The experience left her traumatized and when she was freed from the laboratory she became an ambassador to encourage people to end testing things on animals.

Nan Hauser, a **WHALE** expert, was taken by surprise when she was diving and a humpback whale began to push her around. It wasn't typical behaviour for these gentle giants, and Nan was scared. But she was delighted when she realized that the whale had been trying to protect her from a deadly tiger shark that was lurking nearby.

When a fire broke out in the garage where Scarlett the CAT had left her kittens, she ran back in to rescue them. Scarlett had to face the flames five times before she'd carried her whole litter to safety, and she was badly burned in the process. Thankfully Scarlett recovered and she and her kittens were adopted by well-wishers.

Little Man the LLAMA was just one of Bruce Schumacher's much-loved farm animals. Little Man lived alongside a flock of sheep, and helped Bruce to protect them from coyotes and mountain lions. One day, Bruce was in town when he heard the news that there was a fire on his ranch. He raced back home to discover that his house and barn had burned down, and most of his animals had died. But he found 30 sheep unharmed. Little Man had herded them away from the fire. However, the llama had suffered from burns, and the smoke had damaged his lungs. Sadly, Little Man didn't survive, but Bruce was forever grateful to his llama for being so courageous.

When a marmot fell into a water trough at an American zoo it became the star of a rescue effort – not by a human worker, but by a big ELK called Shooter. The marmot couldn't escape and was in danger of drowning, but the huge male deer gently lifted it out with his teeth and laid it on the ground to dry.

CONGO
— THE —
ARTISTIC CHIMP

The auctioneer's hammer went down with a bang and, in 2005, art history was made. Three paintings had sold for the princely sum of £14,400, but they were no ordinary pieces of art. They were painted by Congo, a chimpanzee.

Congo was born in the wild, but lived at London Zoo and in the 1950s he often appeared on a television programme called *Zoo Time*. The show was presented by Desmond Morris, who was well known as a painter and zoologist as well as a television celebrity.

When Congo was 18 months old Desmond gave him some card and pencils, curious to see what one of our ape cousins might make of these simple tools. To his surprise, Congo drew a series of lines. Over the next two years, Congo experimented with paint and produced about 400 pieces of art. Some of them were even exhibited at the Institute of Contemporary Arts in London.

Art critics argued over whether the chimp's paintings were true art; some claimed they were just random daubs of paint. However, creative geniuses, including Pablo Picasso and Salvador Dalí, admired the paintings and noted that Congo chose colours with care and preferred certain patterns, especially fan shapes. Congo would even throw a tantrum if a painting was removed before he had finished it!

Congo died in 1964. We can never know what a chimp is thinking when it paints, but Congo's paintings may have helped us begin to understand how our own ancestors first experimented with art.

FAMOUS CHIMPS

David Greybeard

Jane Goodall studied wild chimpanzees in Gombe National Park in Tanzania, where she could observe them behaving naturally. One of her favourites was a chimp she named DAVID GREYBEARD. He was the first chimp ever seen to use tools and eat meat. It was ground-breaking research, and assured David Greybeard a place in the natural history books.

Washoe Talks

The first animal to be taught sign language was a chimp called WASHOE, in the 1960s. Washoe could use about 130 signs, but she understood many more. She was eventually allowed to live with other chimps, and she taught sign language to a young chimp called Loulis. Although scientists accept that chimps could use signs, they did not all agree that this meant the chimps were actually 'speaking' a language, like humans talking to each other. Washoe died in 2007, and her life and skills helped humans to understand just how clever chimps are.

Chimps seem human-like in many ways, so it's no wonder that people have kept chimps in captivity to study them or as pets. Today, we know that chimps belong in the wild, with their own families.

Clever Kanzi

KANZI was a bonobo who was able to understand simple instructions that humans said to him, such as 'Put the keys in the refrigerator'. He could 'talk' back to people by pressing symbols on a keyboard.

ANIMALS AND ART

Some animals that can hold a paintbrush have been trained to daub paint on paper, and some people think this activity can reduce boredom for captive animals. No one knows whether the animals have any concept of using colour in a creative way, or take any pleasure from looking at the final product. Some animal paintings are sold to raise money for the sanctuaries, aquariums and wildlife parks where the 'artists' live.

Sea Lions

There is no doubt that sea lions are intelligent and enjoy learning new things. Jay was a sea lion that lived in an aquarium in Japan. He learned how to paint Chinese characters, and his painting performances were very popular with visitors. Morgan is a sea lion that paints in his British aquarium, often choosing his favourite colours of red and orange. Lea was a sea lion in an aquarium in the US, where she would dab her flippers in paint and then press them against paper to make unique 'flipper prints'.

Elephants

An elephant's trunk is almost as agile as a hand, so an Asian elephant with a paintbrush is capable of creating fine paint strokes. However, it takes a great deal of training to teach an elephant to produce a painting, as it's not natural behaviour for them.

Pigeons

Some animals can recognize art created by some well-known artists. Pigeons have been trained to distinguish a Monet painting from one created by Picasso, and can decide whether paintings produced by other famous artists were from the Cubist movement or were examples of Impressionism.

Nature's Artists

Wild animals often create beautiful patterns as part of their normal behaviour. As snakes sidewind across hot desert sand, they leave distinctive zig zag patterns, and honeybees create intricate hexagons when they build wax cells to store eggs and honey. Male bowerbirds collect colourful objects to decorate their bowers, where they dance and strut to impress female birds.

CAN ANIMALS USE TOOLS?

Some animals, whether they have hands or not, are able to use tools. Apes and monkeys use their hands to grip and hold like humans do, but others use different body parts, such as mouths and noses.

CAPUCHIN MONKEYS hold heavy rocks and use them to smash open nuts.

Furry **SEA OTTERS** use rocks like hammers to knock shellfish from rocks. They also float on their backs and rest rocks on their bellies, and wallop shellfish onto the stones until they crack open.

ORANGUTANS wrap up prickly fruit in leaves so they don't get sore hands when they prise the fruit open.

Figaro is a **COCKATOO** who made a tool from wood and used it to reach nuts that were outside his cage. Other cockatoos watched Figaro and soon they were copying him.

BOTTLENOSE DOLPHINS in Australia use marine sponges as tools. They balance the sponges on the tips of their delicate snouts for protection while they forage for food on the seabed. One dolphin began the trend, and the other dolphins learned how to use sponges by watching her.

A wild **CHIMP** was seen using a long stick to reach a banana that was floating in a river.

CROWS in Japan are super-smart tool-users. They wait until traffic lights turn red before dropping nuts on a road. Once cars have driven over the nuts, and crushed them, the birds swoop down to retrieve their food.

Leah, a wild **GORILLA**, used a branch as a walking stick to help her cross a deep river.

PRIMATES IN DANGER

More than half of all species of primate are threatened with extinction and 88 species are critically endangered, which means they are likely to become extinct soon. For most primates, the loss of their forest homes is the biggest threat to their future.

A third species of orangutan, the Tapanuli orangutan, was only identified in 2017. There are fewer than 800 of them in the wild, making this the rarest of all great ape species. Orangutans' homes have been turned into farmland, especially palm oil plantations.

In 1973, there were about 290,000 orangutans in Borneo. Scientists estimate that today there are approximately 50,000 left. That's a catastrophic loss in just over 50 years – the lifespan of one orangutan.

There are fewer than 50 mature northern sportive lemurs left alive. They are likely to be extinct soon unless immediate conservation work can save their habitat in Madagascar.

A hundred years ago there were about two million chimps. Today, there are probably fewer than 300,000 of them in the wild. The main threat to these great apes is cutting down their forest homes to dig mines, build homes for humans and to farm the land.

And some good news! Thirty years of conservation work has helped to save golden lion tamarins from the brink of extinction. In the 1990s, just 272 of these beautiful little primates were counted. The most recent estimates suggest there are now about 1,000 of them.

THERE ARE ABOUT 14,000 SUMATRAN ORANGUTANS IN THE WILD.

ALL TYPES OF GORILLA ARE ENDANGERED, BUT THE CROSS RIVER GORILLA IS A SUBSPECIES AT PARTICULAR RISK. THERE ARE NO MORE THAN 250 MATURE ADULTS ALIVE.

MACHLI
— THE —
CROCODILE KILLER

A mother's love for her children, and an overwhelming instinct to protect them, are powerful passions. Together they help explain the extraordinary courage of a Bengal tiger named Machli.

As soon as she was born, in the mid 1990s, the Indian guides at Ranthambore National Park knew this little cub was different. With a curious nature, and an unusual fish-shaped mark on her face, she was always easy to spot. They called her Machli, which means 'fish' in Hindi, but they could never have predicted that this 'fish' would one day kill a crocodile.

As she grew, Machli showed none of the timidity that most young animals display. By the age of two she had already begun hunting alone, and her speed and skill meant she could leave her mother's care to establish her own territory and family.

By 2003 Machli was renowned as the Queen of the Ranthambore Lakes – crowned by the guides as a fearless mother and hunter. Machli's spirited nature showed no sign of diminishing with age; she was hugely popular with tourists, especially when she brazenly used their vehicles to hide behind while stalking sambar deer! Machli was raising her second set of cubs – twins Jhumri and Jhumru – in the stunning setting of the lakes, overlooked by a majestic palace and fort. Machli's family, however, faced a new and perilous threat: crocodiles.

These powerful predators had always lived in the park, but their numbers had swelled to almost a hundred. The deer had fled the area, leaving the tigers hungry

and fearful. Not only had their food disappeared, but the huge crocodiles posed a threat to their cubs. Only the bravest tiger would ever attempt to battle a crocodile … and that tiger was Machli.

Face to face with a 4-metre-long crocodile, most tigers might snarl then flee, but Machli was no coward and she had confidence in her capacity to overcome a mighty foe – even one twice her size. Mindful of her cubs nearby, she leapt onto the crocodile's back, gripping its massive body with her claws. With agility and speed she avoided the crocodile's long, tooth-lined jaws and swivelled its body round so she could kill it with one bite to the back of the neck.

It was a victory that changed Machli's life forever. The attack had been filmed by tourists, and news of the tiger's astounding feat raced around the world, drawing more visitors and scientists to the park. Machli became one of the most photographed tigers that had ever lived and her fame helped to promote tiger conservation. She had 11 cubs in total, and she continued to prove her courage as she protected them from the many threats that came their way.

DEADLY KILLERS

What is the Deadliest Shark?

Shark attacks are extremely rare, but when people are chased by sharks they don't always have time to try and work out which species of shark is in pursuit! However, the mackerel and requiem groups of sharks contain many deadly types, including the **GREAT WHITE SHARK**. This giant beast is not interested in eating human beings, but it does sometimes mistake people for its favourite snack of seals. The tiger shark, however, is much less fussy and will eat almost anything.

Do Man-Eating Tigers Really Exist?

Yes – sadly tigers are not fussy and are not especially fearful of humans, so sometimes they do hunt men, women and children who enter their territory. The most famous man-eating tiger of all time was the **CHAMPAWAT TIGER**. She is believed to have killed more humans than any other single animal. She is credited with the deaths of 200 people in Nepal and 236 people in India before being put down in 1907. Today, tigers are very rare and people have already caused the extinction of three types of tiger.

Hidden Hippos

Hippos are plant-eaters, but they are regarded as some of the most dangerous animals in Africa. When the Sun goes down, heavy hippopotamuses climb out of their watery home and begin to graze on plants by the river's edge. As the sun rises, they return to the water and that's when these huge herbivores can become especially deadly. Mothers and their calves love to wallow and sink below the surface, but if boats get too close a mother can turn violent to protect her calf. Hippos have huge jaws and enormous teeth, and they can easily capsize a boat or kill a person with a single bite.

TIGERS KILL FEWER PEOPLE THAN LIONS.

BUT LEOPARDS KILL MORE PEOPLE THAN EITHER LIONS OR TIGERS.

TOP ANIMAL PARENTS

The animal kingdom is full of creatures that are brilliant parents to their young.

A mother **OCTOPUS** can produce 100,000 eggs at a time. She lays them in a den and then sits over them, wafting over clean water with her eight arms, and protecting the eggs from predators. She won't leave them to find food, so while they grow the mother doesn't eat. Eventually, the eggs hatch and the baby octopuses swim away, but the starving mother dies.

A female jawfish lays her eggs in her mate's mouth, where he keeps them while they grow.

Male **SEAHORSES** look after the eggs, not the females. A father keep the eggs in a pouch on his belly until they are ready to hatch.

A **SURINAM TOAD** keeps about 100 eggs in little pockets that develop on the skin of her back. The eggs can grow safely there, until they are ready to hatch and tiny wriggly tadpoles break out.

KANGAROOS, KOALAS and other marsupials give birth to tiny babies that are no bigger than jellybeans. Each baby is called a joey and it crawls into a pouch on its mother's belly and lives there while it grows, feeding on her milk.

A male EMU builds a nest where the female lays her eggs. Dad stays with the eggs for eight weeks, turning them several times a day so they develop properly. When the chicks hatch he will look after them and teach them how to find food.

An elephant mother is pregnant for **22 MONTHS** before she gives birth.

CUCKOOS have an unusual approach to parenting. They lay their eggs in the nests of other birds and let those birds do all the hard work. When the cuckoo chicks hatch they push any non-cuckoo chicks out of the nest, so they get all the love and attention for themselves.

SEA OTTERS are now protected after their numbers fell dramatically. Wildlife experts in California wanted to make sure that every sea otter had a good chance of survival, so they recruited two female sea otters to become foster mums. When orphaned sea otters arrived at the wildlife sanctuary, they were given to Toola or Joy, two motherly sea otters who loved to take in new 'babies' and care for them.

Did you know?

Some animals, such as many fish and bugs, have huge numbers of offspring at a time. They produce thousands, or even hundreds of thousands, of young but they don't invest time and energy in looking after them. Other animals, such as cats, have fewer offspring, but they do take care of them, sometimes for years. Each way of parenting can be successful in ensuring that a species survives.

GI JOE SAVES THE DAY

In August 1946, a pigeon named GI Joe was honoured for saving the lives of at least 100 soldiers. He was given the Dickin Medal, a special honour for animals that have performed acts of outstanding bravery or devotion to duty at a time of war.

Today we take modern communications for granted, but before the invention of satellites and mobile phones people still used some ancient ways of passing messages to one another. Trained homing pigeons have long proved a fast and reliable way to reach allies and comrades. One of the first recorded uses of these intelligent birds was more than 2,000 years ago, when the great Roman commander Julius Caesar conquered Gaul – modern-day France.

During the Second World War, homing pigeons were in much demand by all sides. In the United Kingdom alone an estimated 200,000 birds were used to help the war effort. They would be taken to fields of war and later released, with messages tied to their legs, to return to the places they had learned were 'home'. American, German and French troops also benefitted from pigeons' incredible ability to fly great distances and find their way back through wind, snow and even gunfire.

GI Joe was an American homing pigeon who began his service in 1943 as part of the United States Army Pigeon Service. The German Army had occupied the Italian town of Calvi Vecchia and the Americans were putting the final touches to a plan for an air raid to help British troops remove the enemy. However, British troops had already launched their attack and they had successfully liberated the town earlier than expected.

The crewmen of the Allied XII Air Support Command were due to bomb Calvi Vecchia imminently, but the British troops were now stationed in the line of fire. Efforts to send a radio message to call off the bombing raid failed. Disaster loomed – there were less than 30 minutes before the town, its inhabitants and the British troops would be hit with the full force of a deadly bombing raid.

With no time to waste, GI Joe was despatched with a message. He was released into the air, carrying with him the desperate hopes of many brave soldiers. The pigeon flew at great speed, covering 32 kilometres in 20 minutes, and reached the Allied air force just as the planes prepared for take off. If he had arrived just five minutes later the bombs would have fallen, costing the lives of at least 100 servicemen and causing untold carnage in the town.

GI Joe was given his Dickin Medal for gallantry at the Tower of London and the citation read 'This bird is credited with making the most outstanding flight by a USA Army Pigeon in World War II'. He was the first non-British animal to receive the award.

After the war, GI Joe was returned, along with another 24 heroic pigeons, to the US Army's Pigeon Breeding and Training Centre in Fort Monmouth, New Jersey. A total of 32 pigeons were awarded the Dickin Medal before the use of pigeons in war finally came to an end.

The United States Army Pigeon Service was shut down in 1957, and GI Joe was cared for in a zoo until his death in 1961, aged 18.

HOW DO ANIMALS FIND THEIR WAY?

Many birds are superb NAVIGATORS. Some scientists think they use a number of clues, including an ability to sense the Earth's magnetic field. This would enable them to identify where north is, and use that information to guide their route. Birds may also learn routes over the course of several migrations, and younger birds in the flock follow them.

WILDEBEEST in Africa travel across the grasslands to reach breeding and feeding grounds, but no one knows how they find their way as their routes vary from year to year. They may look for rainclouds in the distance and head towards them.

MONARCH BUTTERFLIES weigh less than a bean, but can undertake incredible journeys, flying 45 kilometres a day. They swarm in vast numbers, heading south to spend the winter in warm Mexican or Californian forests. The butterflies use many clues to help them to find their way, including the Sun, magnetic fields and smells. They may also use landmarks, such as the Rocky Mountains, to guide them.

WANDERING GLIDERS

are dragonflies that travel great distances across oceans. Some of them regularly appear in North America, but no one knows where they travelled from, where they go or how they know which way to go.

SALMON spend most of their lives in the sea, but they swim to rivers where they mate and lay their eggs. The young salmon then swim back to the sea. Recent research suggests that salmon can find their way to rivers because they have a magnetic 'map' and they can sense tiny local changes in the Earth's magnetic field.

INCREDIBLE JOURNEYS

A migration is a journey that animals embark on to find better conditions or resources, such as food, water and mates. From birds to sealife, some animals experience the most amazing expeditions in their lifetimes.

On the Wing

When a little ARCTIC TERN hatches from its egg it cannot begin to imagine the life it has ahead. It will travel the world, see more places and fly further than any of us could ever dream possible. This little bird, which starts its life in Greenland near the North Pole, will soon embark on the first of many flights to Antarctica and back again. During its lifetime, an Arctic tern will fly the equivalent distance of three round trips to the Moon. When a strong wind is behind them, Arctic terns can fly up to 670 kilometres in a day.

The Mysterious Eel

EUROPEAN EELS endure one of the longest and strangest migrations. Scientists have discovered that eels live in rivers in Europe, then swim downstream to reach the Atlantic Ocean and then trek 5,000 kilometres to reach the Sargasso Sea, in the North Atlantic Ocean. Once there, they mate, lay their eggs and die. It takes the baby fish a year to swim all the way back to Europe again, and most of them die long before they will be able to return to the Sargasso Sea.

Astonishingly, if EUROPEAN EELS hatch in eel farms in Europe, they can still find their way to the Sargasso Sea. No one knows how they do it.

Intrepid Turtles

It's evening and the Sun hangs heavy in the air, its golden light reflecting off a sapphire-blue sea as it sinks below the horizon. This is the cue for hundreds of baby turtles to make their move. Guided by the dying light, or a rising Moon, the babies are facing a life-or-death race to the glistening water. As they scramble across the sand, however, birds, dogs and crabs appear from nowhere, ready to pick off any juicy reptile snacks they can grab.

They eventually reach the lapping water, but the baby turtles are tossed around in the froth as the waves hit the beach. They are thrown back onto the sand, but the turtles don't give up. They right themselves and once again head for the water. The survivors dive beneath the surface and prepare to spend a life at sea. When a female is mature enough to breed, she must make her way back to the beach where she hatched. This is where she will lay her eggs, burying them in the sand to keep them cool and safe from predators.

DID YOU KNOW?

During its lifetime, a **LEATHERBACK TURTLE** swims thousands of kilometres through the ocean, helped by currents to make long migrations to find food and mates. One leatherback turtle swam 20,557 kilometres on its migration between Indonesia and America.

Courageous Crabs

There are up to 120 million **RED CRABS** living on Christmas Island, south of Indonesia. Every year, between October and December, crabs leave their rainforest home and head towards the sea, where females will release their eggs. They march in a straight line, crossing roads and climbing cliffs to reach their destination.

BEAU AND BEATRICE

Farmer Donald MacIntyre walked through the crisp January air. It was early morning and he was planning to check up on his 23 shire horses. But as he entered the stables, Donald's heart sank.

Beatrice, a 1-tonne mare, was lying on the stable floor in great pain. Donald knew he had to get the heavy horse back on her feet, so he called his partner, Jane, and four farm hands to the stables. No matter how hard they tried they could not raise the horse.

Beatrice's agony was caused by equine colic and if she stayed on the ground she was likely to develop organ failure. For six long hours she lay on her side as everyone tried to raise her up with a tractor and straps, but she couldn't be budged. Beatrice's temperature was reaching dangerous levels and her heart rate dropped. Donald knew that his beautiful mare was close to death and decided to let her spend her last few minutes at peace.

Beau, Beatrice's mate, had been watching the attempts to save her from his stall. Donald brought him over to say goodbye to Beatrice, but instead Beau nipped at the mare's ears and neck, and grabbed hold of her halter with his teeth. To everyone's astonishment, Beau lifted up his mare's head and helped the trembling horse get to her feet.

Thanks to Beau's perseverance and determination, Beatrice made a full recovery and a month later Donald discovered she was in foal. A healthy colt, Angus, was born in March 2018.

HEROIC HORSES

Sefton

In 1982, tragedy struck in London, UK, when a bomb was planted and timed to blow up as 16 soldiers and their horses were making their way to the Changing of the Guard. Four soldiers were killed by the bomb and seven horses died too. All nine of the remaining horses were injured, including Sefton. He was a large black horse who had been born in Ireland in 1963. He had a distinctive white blaze on his face and four white socks, but he was mostly known for his bold and sometimes stubborn character.

Sefton was badly hurt. People ran to help the wounded soldiers and tend to the injured horses. One man used his shirt to stem the flow of blood from Sefton's neck and he was soon put into a horsebox and taken away for emergency medical care. He required urgent surgery which lasted eight hours, but Sefton slowly recovered alongside another injured horse, Echo, and they became firm friends. Echo retired from duty, but soon Sefton was back at work and hailed as a national hero. Eventually, Sefton joined his old friend Echo in retirement, and he died in 1993.

Sunny Boy

Horses are able to form intense bonds with humans because they are herd animals, and often see 'their' humans as part of the herd. The strength of these bonds was evident in 2008, when Chloe-Jeane Wendell and her sister Kristen were riding their horses in a festival parade in Louisiana, USA.

Suddenly, a pit-bull dog raced out from the crowd and attacked Kristen's horse, Angel. The horse kicked back, so Kristen leapt off her mount to avoid being thrown, and the dog went to attack her instead. Chloe-Jeane quickly jumped down from Sunny Boy to rescue her sister, but brave Sunny Boy took control of the situation instead. He fearlessly placed himself in front of the girls and kicked the dog in the face, until it moved away. The dog was caught and the two girls were in no doubt that Sunny Boy had saved them from what could have been a fatal attack.

Kerry Gold

Farmers know that cattle can be dangerous beasts. When cows and bulls are frightened they may attack the people or dogs who threaten them.

Dairy farmer Fiona Boyd was caring for a calf on her farm in Scotland. The calf's mother mistakenly thought her offspring was in danger, and attacked Fiona, sending her flying to the ground. She stood over the woman menacingly. Every time Fiona tried to get up, the cow knocked her to the floor again. With one well-placed step, the cow could have killed Fiona, so she curled herself into a ball to protect herself.

Fiona realized that escape was impossible and she lay on the ground, terrified, when suddenly the cow began to move away from her huddled body. Kerry Gold, Fiona's chestnut Arab mare, had seen what was happening and had begun to neigh and snort. She galloped over to the cow and began kicking it. The cow retreated and Kerry Gold stood guard while Fiona made her escape. She rang her husband, who was in a nearby field, and he drove Fiona to hospital. Fiona had escaped with just a few bruises, but afterwards Kerry Gold took on the job of Fiona's bodyguard, and followed Fiona whenever she visited her cows. They were always a little bit scared of the horse after that day, and kept a safe distance.

HORSES
DID YOU KNOW?

WHEN DID PEOPLE FIRST RIDE HORSES?

In prehistoric times horses were hunted for food, but by about 3000 BCE wild horses were being domesticated. It's possible that the first people to keep horses were the Botai people of Central Asia, who kept them to ride, and for their milk and meat.

WHEN WERE HORSES FIRST USED IN WAR?

As far back as 4,500 years ago the Sumerians used war chariots that were pulled by donkeys or asses, but by 1600 BCE horses were used instead. Archaeologists have found a manual for training horses to pull chariots that was written in about 1350 BCE. Cavalry horses are ones that are ridden by soldiers, and they were often used by ancient armies, including those of the Roman and Greek civilizations. By the Middle Ages, the cavalry was an important part of warfare in Asia and Europe.

HOW MANY DIFFERENT TYPES OF HORSE ARE THERE?

There are hundreds of breeds of horse, but they are divided into three main types:

1. **COLDBLOOD HORSES** are huge, heavy and strong. They are particularly useful for pulling hefty loads and in the Middle Ages they were bred to carry the great weight of knights in armour.

2. **WARMBLOOD HORSES** are popular for riding as they have mild temperaments. They can be trained for sport and they make good pets. Most warmblood horse breeds were developed in Europe.

3. **HOTBLOOD HORSES** are the fastest, most lively and spirited horses. They include the Arabian and thoroughbred breeds. The Arabian is thought to be one of the oldest horse breeds in the world.

SMOKY TAKES TO THE SKIES

Small dogs are famous for being fearless, loyal and full of energy. Those are just three of the many fine qualities that saved Smoky the Yorkshire terrier on more than one occasion. She may have been petite, but this dog had an enormous personality and she put it to good use.

Smoky's incredible story began in 1944, but mystery still surrounds her early days. American soldiers found her on the tropical island of New Guinea, sheltering in a hole near a road not far from lush rainforests. It was a strange place to find a lovable lapdog, and no one ever discovered how she got there. Whatever the cause of her predicament, the little dog was lost, she was hungry and her fur was a scrawny mess.

Corporal Bill Wynne took Smoky into his care, and with his love and attention she quickly transformed into a bright-eyed pet and companion. She slept in Bill's tent, shared his rations and even sat in his backpack when he went on combat missions. Smoky survived air raids and hurricane-strength tropical storms and Bill claimed she even saved his life when she alerted him to incoming shells on a transport ship.

Smoky was unusually clever and needed plenty of stimulation, so Bill taught her tricks. She learned to ride a scooter, walk a tightrope and she even jumped from a plane seven times, fully equipped with her own little parachute! On the last occasion the wind suddenly dropped as Smoky was falling to the ground, and her parachute collapsed. Smoky was blown far from the catching blanket. Although she landed safely, Bill decided that it was one jump too many!

When Bill fell ill with dengue fever – a deadly tropical disease – he was sent to hospital for treatment. His friends brought Smoky to visit him and she quickly won over the medical staff who let her spend time with wounded soldiers. Her joyful character and playful nature had a strong effect on the men. They laughed as they watched her race around the ward, chasing enormous birdwing butterflies.

Smoky wasn't just a delightful dog that brought joy to Bill and his comrades. She really proved her worth when she helped the war effort more directly. A new airfield for allied planes was being constructed and engineers needed to install telegraph wires through a pipe to reach the airfield. The problem was the pipe was more than 20 metres long and only 20 centimetres wide, and it was partly blocked by soil in many places.

The obvious solution was to dig a new trench and lay the wire by hand, but that would leave soldiers exposed to enemy bombing for three days. Bill had an idea. He attached the wire to Smoky's collar and sent her through the pipe. Following Bill's gentle coaxing, Smoky raced through the pipe and emerged from the darkness in a cloud of dust. She and Bill had ensured the success of the communications network in minutes.

Smoky was credited with possibly saving as many as 250 lives and 40 planes that day. After the war she continued to visit soldiers in hospital and helped them to recover both physically and mentally. She retired in 1955 and died peacefully in her sleep two years later.

OUR BEST FRIENDS

Everyone knows that an animal can provide love, company and friendship as a pet. Sometimes an animal goes one step further and becomes a professional comfort animal.

Dan McManus suffers from anxiety but his dog **SHADOW**, an **AUSTRALIAN CATTLE DOG**, helps him to cope. He also likes to go hang-gliding, but Shadow didn't like Dan going without him, and would even bite onto Dan's boots to stop him from getting in the air. Eventually, Dan had a special harness made for Shadow, so that his faithful friend could go hang-gliding with him.

KOSHKA was a **STRAY CAT** who became the beloved companion of an American soldier called Jesse who was serving in Afghanistan. There were dark days for Jesse, when friends and colleagues were killed, and Jesse wasn't sure he could handle the emotional strain of war. But Koshka always seemed to know when Jesse needed extra love and attention, and Jesse says that Koshka saved his life. He repaid his friend by arranging for Koshka to travel to the United States, where he now lives with Jesse's parents.

WILLOW AND WINNIE are **MINIATURE HORSES** that work as comfort animals. They visit people in nursing homes, hospitals and homes for severely disabled people. These two mini-horses are no bigger than a large dog, so they can easily move around indoors, and people love to stroke them.

ALTHOUGH DOGS AND CATS ARE THE MOST COMMON COMFORT ANIMALS, SOME PEOPLE SAY THEY GET EMOTIONAL SUPPORT FROM MORE UNUSUAL ANIMALS, INCLUDING FERRETS, GOATS, PIGS, MICE AND BIRDS.

WAR DOG STORIES

LUCCA was born in the Netherlands but trained with the United States Marine Corps as a search dog to sniff out explosives. She was involved in more than 400 missions in Iraq and Afghanistan, and helped to save many lives. On her last mission in 2012, Lucca discovered a huge IED – an improvised explosive device – but while she was looking for others, a second device detonated. Lucca lost a leg in the explosion and suffered severe burns. She was retired from service and received a medal for her 'gallantry and devotion to duty'.

During the Second World War, **IRMA** had an important job to do in London. She was tasked with finding people buried in the rubble of their homes after bombs had landed on them. She saved hundreds of people, including two young girls who were trapped under a collapsed house. Irma had refused to move until they were rescued.

BAMSE served at sea with the Free Norwegian Forces during the Second World War. One day, he saved a crewmate who had fallen overboard by dragging him back to the shore.

REX AND THORN were particularly brave dogs of the Second World War. Every animal has a natural fear of fire, but these courageous Alsatians had to enter burning buildings to help firefighters find people trapped within. They faced choking smoke and flames to do their work, and saved many lives.

During the Second World War, many men were held captive in Japanese prisoner-of-war camps. In one camp there was an English pointer, called **JUDY**, who had been on a British ship that had been bombed. She lived with her fellow prisoners and often risked her own life by fearlessly attacking the guards when they beat her human friends.

CHIPS fought on the front line in the Second World War, travelling to Germany, France and North Africa. One of his bravest actions occurred on a beach when he and his colleagues were facing relentless machine-gun fire. Chips broke free from his handler, and ran at the hidden gun nest. He grabbed the enemy soldier who was operating the gun and dragged him out.

STUBBY the pit-bull terrier served in the First World War. He was the first dog to earn rank in the United States Armed Forces, and was made a Sergeant.

MONTAUCIEL
— AND HER —
FOWL FRIENDS

The huge crowd gasped in amazement as the rope cords were cut and the enormous Aerostatic Globe floated smoothly into the air. Inside the wicker basket hanging beneath the globe – which today we would call a hot-air balloon – were a duck, a cockerel and a sheep called Montauciel. Their flight was about to head straight into the history books.

The spectators, including King Louis XVI of France and Queen Marie Antoinette, had gathered at the royal Palace of Versailles in September 1783 to witness the first 'manned' air flight. The balloon was constructed from cloth and beautifully decorated thanks to the designs of a wallpaper manufacturer.

The balloon's nervous inventors, brothers Joseph and Étienne Montgolfier, watched closely as the fire beneath the globe heated the air inside it, and slowly lifted the huge blue-and-golden balloon skywards. It was carried by a brisk south-westerly wind and floated for about eight minutes before landing safely in a forest nearby. The crowds chased after the balloon and were delighted to find that the birds, while a little dazed by the experience, were unharmed and Montauciel was happily grazing close to the overturned wicker basket.

Having seen that flight was not dangerous for a sheep, the King agreed to let the Montgolfier brothers experiment with humans. They set to work preparing for their first human flight in a hot-air balloon just a few weeks later. Thanks to Montauciel, the flying sheep, the science of aviation was ready to take off.

ANIMALS IN SPACE

The first animals to be blasted into space were **FRUIT FLIES**. In 1947, they were launched from the US on board a former Nazi V-2 rocket and flew 108 kilometres into the air and returned to Earth safely with the help of a parachute.

The journey from Earth to space is phenomenally dangerous, and in the early days of space travel no one really knew what the effect of such journeys might be on the human body. Animals were rocketed into space first to test the equipment and conditions, and establish the chances of survival for human explorers.

It's no wonder that **MONKEYS AND APES** were among the first animals sent into space, as they are most similar to humans. There have been 32 sent to date. The first monkeys to survive a space flight and return to Earth were called Able and Baker. Their space flight took place in 1959, the same year that a rabbit went into space.

During the 1950s, the Soviets chose **12 STRAY DOGS** as their animal-astronauts. A dog named Laika became the first living being to orbit the Earth, on board Sputnik 2 in 1957. Sadly, she died from heat and stress after just a few hours. In 1960, the Soviet rocket Sputnik 5 had a menagerie of passengers – two dogs, a rabbit, 42 mice, two rats and some fruit flies.

In January 1961, **HAM THE CHIMP** became the first chimpanzee in space. His mission was a success and his safe return to Earth was an important step towards human space flights. Just three months later, **YURI GAGARIN** became the first person to fly in space, orbiting the Earth in a Vostok spacecraft.

By the late 1960s, the Americans and Russians were competing to see who could get a human on the Moon first. In 1968, the Soviet Union sent the Zond 5 rocket to orbit the Moon, carrying **FLIES, MEALWORMS, PLANTS** and two **TORTOISES**. In 1969, humans first landed on the Moon.

Anita and Arabella were the first **SPIDERS** in space. They were put on board NASA's Skylab space station in 1973 so scientists could examine the effect of weightlessness on their web-weaving skills.

Some of the most successful astronauts have been tiny animals called **TARDIGRADES**, or **WATER BEARS**. These invertebrates are able to exist in a desiccated (dried) state, which means they can endure extreme conditions on Earth, so scientists from the European Space Agency were interested to find out how well they would cope with the challenges of space.

In 2007, **DESICCATED WATER BEARS** were included in the Foton-M3 mission that spent 10 days in low-Earth orbit. They returned to Earth safely and were rehydrated. The water bears were discovered to have survived the airless conditions, cosmic rays and the space vacuum, and some of them were even able to lay eggs. Their desiccated state allowed them to withstand some of the worst effects of solar radiation.

WATER BEARS are the only animals to survive open space.

MEDIA STARS

Many animals have achieved celebrity status over the centuries. For some animals, fame isn't the result of a one-off event – it's a full-time job.

Psychic Stars

PAUL the octopus was the unlikely star of the 2010 World Cup. He successfully predicted the outcome of eight matches, by choosing a mussel from one of two boxes bearing the flags of the competing countries. As the tournament went on, Paul's psychic talent drew huge attention from the world's media and football fans.

Internet Influencers

GRUMPY CAT, whose real name was Tardar Sauce, became an overnight Internet sensation after a photo of her glum face was posted online in 2012. She appeared on TV shows and starred in her own video game. Grumpy Cat had many millions of followers on social media and she even had her own line of merchandise!

Animal Actors

One of the most famous animal actors of all time was a dog called RIN TIN TIN. He began his life in France, during the First World War, but he was rescued by an American soldier called Lee Duncan after his kennel was bombed. After the war, Lee took the black German Shepherd back to his home in Los Angeles and together they competed in a dog show, where Rin Tin Tin's talents were spotted. His first major film appearance, in 1923, was an instant hit and he went on to star in more than 20 films.

When animals are used in films, the company that is making the film is expected to follow guidelines that ensure the animals are not harmed. Safety staff monitor the work that animals are expected to do and report on their welfare.

NATURE'S LAB

Scientists, engineers and inventors are inspired by the natural world to find solutions for human problems.

Swimsuits

The SKIN OF A SHARK is covered in tiny tooth-like scales that are made from the same tough substance as our teeth. These denticles are shaped and arranged in such a way that water flows smoothly over them, helping the sharks to swim fast. Scientists have copied shark skin to create high-performance swimsuits.

Bullet Trains

SUPERFAST TRAINS in Japan are called 'bullet trains' and were modelled on a bullet shape to increase streamlining, but they made an unpleasant booming sound when they left a tunnel. Engineers noticed that kingfisher birds produced very little splash as they entered the water to catch fish. They used the shape of a kingfisher's head and beak to help them to design new bullet trains, and discovered that the trains were more aerodynamic and that noisy boom disappeared.

Colour Screens

The colours on a BUTTERFLY'S WINGS are partly due to the colour of the microscopic scales that cover the wings, but also to the ridged structure of those scales, which reflect and refract light to produce a metallic shimmering effect called iridescence. Scientists hope to copy this combination of colour and structure to create display screens that use less energy than current technology.

Wind Turbines

HUMPBACK WHALES are superb swimmers, but they have strange bumps on their fins. Scientists discovered that these bumps enable the whales to turn tight corners and manoeuvre their huge bodies with ease. When bumps were added to the blades of wind turbines, scientists discovered a similar effect, making them more efficient and quieter.

A LION — CALLED — CHRISTIAN

It was a year since John Rendall and Ace Bourke had seen Christian, and they feared that their journey to see him again, during which they had travelled thousands of miles, would be in vain.

With legendary lion expert George Adamson by their side, the young Australians were scouring the dry African landscape for any sign of this special lion. They walked for hours but there was no sign of Christian and, although they were disappointed, they knew that the young male was living free, as he should.

Suddenly, a bouncing flash of golden fur bowled up towards them – it was Christian, heading at full speed towards John and Ace. Instead of flinching or taking cover, John and Ace smiled and opened their arms wide as Christian fell on to them both, embracing the men as they hugged him back. The best friends were reunited!

John and Ace had first met Christian when he was a tiny cub in London's famous Harrods department store. In 1969, keeping exotic pets was fashionable and legal, and the Harrods pet department often supplied rare and unusual pets to anyone who could prove they would care for them.

As soon as the men saw Christian they were entranced by his personality and decided to buy him. The two friends were working in a furniture shop, and living in a flat above it. They installed Christian in the shop's large basement where he had plenty of room to play, hide and run. They took him to a park for daily exercise, so that Christian could practise his chasing and pouncing skills.

Like all cubs, Christian grew quickly and John and Ace began to worry about his future. While they felt safe with the lion, they knew that Christian would soon be strong enough to overpower them. It was also obvious that London was no place for a lion, but they were desperate to save Christian from life in a zoo.

One day, film actors Bill Travers and Virginia McKenna came to the furniture shop. They had starred in a hugely successful film, *Born Free*, which told the story of Elsa, a lioness who was returned to the wild by wildlife experts George and Joy Adamson. John and Ace introduced Virginia and Bill to Christian, and asked for their advice. It wasn't long before everyone agreed that Christian should be prepared for a journey to Africa, where George Adamson could help him to adjust to life as a wild predator.

In 1970, Christian left for George's lion reserve in Kenya. John and Ace said goodbye with heavy hearts, but knew they were giving Christian the best chance of a happy future. Despite many difficulties, the experiment proved to be a great success and Christian soon adapted to his new life. He began to hunt for his own food and eventually found a mate and began his own pride.

John and Ace were reunited twice with Christian before he finally disappeared into Kenya's deep wilderness. They last saw him in 1973, but they never forgot their beloved friend. In 2006, an old film of their joyful reunion with Christian resurfaced and it was viewed by more than 100 million people. The story of their special bond helped conservationists around the world in their efforts to return captive animals to the wild.

THE FAMILY CATS

LIONS are the only cats that live in family groups called **PRIDES**. Unlike most cats, adult males and females look very different. Males are bigger with a distinctive mane of thick fur around the head and neck.

Each pride is a strong family unit that is usually made up of one or more mature males, up to six females, and their cubs. The males protect the pride from other lions, and father all the cubs. If a younger male successfully challenges the leader of a pride and takes it over he may kill all the cubs so he can quickly mate with the females.

Females are **SUPREME HUNTERS**, working together as a team to find, chase and kill their prey. Their skill in communicating with each other and co-ordinating an attack allows the lions to kill prey larger than themselves.

While females hunt, the male lion looks after the cubs. Young lions enjoy **PLAY-FIGHTING** and it's an important part of learning how to chase and pounce. When youngsters get too boisterous their father may gently wallop them with his huge furry paw.

Most lions live on the grasslands of Africa, but a rare type of lion, the **ASIATIC LION**, is found in the Gir Forest region of India. There are just 350 left in the wild, making them some of the rarest animals alive.

African lions are now extinct in at least 26 countries where they used to thrive.

LION STATS
Size: Up to 2.5 metres long

Lives: Africa and Asia

Habitat: Forests and grasslands

Diet: Animals, including zebras and antelope

BORN FREE, LIVE FREE

Elsa the Lion

The famous wildlife conservationist GEORGE ADAMSON had an important job, but it provided him with some truly terrifying moments. It was 1956 and, as a game warden in Kenya's Northern Frontier District, George knew that deadly wild animals could spring an attack on him at any second. He was prepared, even when a lioness leapt out of the scrub and attacked him. George fired his gun at the predator and killed her. Then he realized that the lioness had only attacked because he was close to a rocky crevice where she had hidden her cubs. She was trying to protect her young and she had paid with her life.

With the help of his wife Joy, George took the three cubs home. The couple arranged for two of them to be sent to a zoo, but they decided that the third, whom they named ELSA, would be returned to the wild. The Adamsons had set themselves a difficult task. They had to rear Elsa by hand, while teaching her how to be a lion and to hunt. Their efforts at rehabilitation were successful, and Elsa became the first lion to be successfully returned to the wild. She had her own litter of cubs, but Elsa died from disease in 1961, her head resting in George's lap as she passed away.

Born Free

After working with Elsa, George decided that he wanted to devote his life to lions. Joy was keen to work with cheetahs, and their careers forced them to live apart. Joy wrote about the experience of living with a lion in her best-selling book, BORN FREE, and it was eventually made into a successful film. Joy died in 1980, and in 1989, the ELSA CONSERVATION TRUST was based at Joy's Kenyan home to teach children about their environment and wildlife.

Kora National Reserve

By 1970, George Adamson had rehabilitated at least 15 lions and he moved to a region in the north of Kenya called Kora to continue his work at a new wildlife reserve. It was at Kora that George was able to release a lion named Boy, who had starred in the *Born Free* film, and Christian the lion who had been bought at Harrods (see page 112). With his assistant Tony Fitzjohn, George established the GEORGE ADAMSON WILDLIFE PRESERVATION TRUST in 1979 to raise funds for their continued work with lions and leopards at Kora National Park. Ten years later, George discovered a female tourist being attacked by poachers. He rushed to her aid, but was shot and killed – along with two of his employees. After he was buried, a wreath of flowers was laid on his grave. It was later dragged away by a lion who had been to visit George's final resting place.

JUAN SALVADOR MAKES FRIENDS

Lying on a beach in Uruguay, coated with oil and struggling to breathe, the little penguin tried to move his flippers. All around him lay hundreds of dead penguins, killed by an oil spill that had polluted their ocean home.

Fortunately for the small bird, a young English teacher called Tom Michell was holidaying near the beach and was taking a walk when he spied the terrible scene of devastation. Tom saw the small movement amongst the lifeless bodies. He scooped up the bird in his arms and took him back to his friends' apartment where he set about cleaning him up.

Juan Salvador, as Tom named the penguin, appeared fit and healthy once the toxic oil was removed, but no matter how many times Tom tried to tempt the penguin back into the ocean, he refused. Juan preferred to stick by Tom's side and repeatedly followed him home from the beach. Eventually, the teacher's holiday was over and he had to admit defeat. Tom returned, with Juan Salvador, to the boarding school in Argentina where he worked.

Juan proved to be very popular with the boys, who enjoyed playing with him in the sports fields and taking turns to feed him with sprats. One 13-year-old boy, Diego, was particularly keen to help look after the penguin. He was very homesick and had few friends. He also struggled to keep up with his schoolwork.

One day, Tom learned that the water in the school pool was due to be changed, so he decided to let Juan Salvador swim there first. If Juan fouled the water, it wouldn't matter – and if the bird refused to get out then he could be retrieved once the pool

was drained. It was to be the penguin's first swim since leaving the ocean, and at first Juan wasn't keen to get his feet wet. With Diego's help, Tom encouraged the bird to dive into the pool, and Juan was soon speeding up and down the pool as the boys cheered him on.

Diego was watching from the side of the pool. He quietly asked Tom if he could join Juan, and dived in almost before Tom had a chance to answer. Within minutes, Diego and Juan were swimming together. As Diego swam, Juan almost danced around him, flying through the murky green water in a figure of eight. Together, the boy and the bird gave a stunning performance of underwater acrobatics. Tom and the other boys watched in awe as Juan and Diego slipped through the water like dancers, their movements perfectly timed to create a breathtaking display of skill and speed.

From that time onwards, Diego was a changed boy. He told Tom that his father had taught him to swim in a river and he gradually began to open up about his life and his feelings. It marked the beginning of a transformation. Diego made friends and his schoolwork began to improve. When the school held its swimming gala, the boy proved just how great his swimming talent was when he won every race he took part in and broke several school swimming records.

Diego was celebrated as a hero, but he knew in his heart that the real hero was his feathered friend, Juan Salvador.

PENGUIN PARENTS
DID YOU KNOW?

WHO ARE NATURE'S MOST HEROIC FATHERS?

An emperor penguin father is one of the most heroic of all animals. He battles with snowstorms, fierce polar winds and near-starvation to rear his chick during the long Antarctic winter.

DO PENGUINS MAKE NESTS?

There are almost no plants in the Antarctic, so emperor penguins can't make normal bird nests. A female lays her egg, and her partner gently rolls it onto his feet and a layer of skin folds over it, keeping it perfectly warm while the air temperature drops below -35°C. Other types of penguin use rocks to make stony nests.

WHERE DO THE MOTHERS GO?

By early June, the emperor penguin females are leaving the colony and heading towards the open ocean to find food. They gorge on fish and fatten themselves up. Meanwhile, the males are left behind with the eggs. They haven't eaten for two months.

HOW DO THE FATHERS SURVIVE?

The Antarctic is the coldest place on Earth, yet emperor penguin males manage to endure more weeks of starvation and extreme weather conditions by using up layers of body fat and huddling together in snowstorms. Eventually, in July, the females head back to find their partners, calling out to them as they trundle past hundreds of other penguins who are all looking forward to a family reunion.

WHO LOOKS AFTER THE CHICKS?

The emperor penguin chicks start to hatch and both parents share the work of keeping their chick warm, and fetching food from the sea. They regurgitate food into their mouths and the chicks eat it from there. By December, the chicks are nearly as tall as their parents and will soon head to the sea.

PENGUIN LIFE

A penguin is a very unusual bird. It has wings that work like flippers to fly through water. It is adapted to life in the cold, with densely packed feathers and a thick layer of body fat for insulation. All penguins live in the southern hemisphere, and most of them are found around the Antarctic and in the Southern Ocean.

Feathered Friends

Penguins live in large groups called colonies, often nesting close to each other and hunting together. African penguins swim in hunting packs, chasing fish into one place until they create a swirling mass of fish called a bait ball. Penguins that work together to catch their food are more than twice as successful as penguins that hunt alone.

Super Swimmers

A penguin's body is torpedo-shaped and it is perched on short legs. It makes the bird appear awkward as it waddles on land, but it's the perfect shape for slicing through water. The fastest swimming bird in the world is the gentoo penguin, with top speeds of 36 kilometres an hour.

Diving Deep

Penguins feed underwater and can stay below the surface for up to 20 minutes at a time. That's quite an achievement for a bird that has to breathe in air. Some penguin species can dive to depths of 400 metres or more as they search for fish to eat.

PENGUIN STATS

Size: Up to 1.1 metres tall

Diet: Fish, squid, krill, jellyfish

Number of species: 18

OCEANS IN PERIL

The oceans cover 70 per cent of the world's surface, making them the largest single habitat available to wildlife, and an essential part of the planet's climate stability and ecosystems. Despite their great importance, humans have shown little respect for the oceans in recent years. Now they are in peril.

Plastic

It has been estimated that the equivalent of one rubbish truck of plastic waste is dumped into the ocean every minute. By 2050, it is predicted that the ocean will contain more plastic, by weight, than fish.

Climate Change

The burning of fossil fuels puts gases in the atmosphere that prevent heat from escaping, causing the planet to warm up. This creates a phenomenon called 'global climate change' and it will probably impact all living things. The oceans and marine life are at particular risk because sea levels are likely to rise and seawater will become more acidic.

Overfishing

Taking too many fish from the sea can cause fish populations to crash. North Atlantic cod, for example, were hunted close to extinction in the last century and bluefin tuna are endangered.

Pollution

Chemicals from farming and factories find their way into rivers and are carried down to the sea. These pollutants, which include deadly metal and chemicals that are used to kill plants and animals, can devastate marine life.

Coral Reefs at Risk

Coral reefs are built by tiny animals, called polyps, that rely on clean, salty, sunlit water to survive. In recent years, reefs have begun to die. They have been affected by warming waters, pollution, and recent evidence suggests that the chemicals in plastics are also killing the polyps.

When reefs die, millions of animals that depend on the habitat are likely to die too. Turtles, sharks, dugongs and whale sharks are just some of the precious animals that could disappear when corals are destroyed.

THE GREAT BARRIER REEF IS HOME TO MORE THAN 5,000 SPECIES OF MOLLUSCS, 1,500 SPECIES OF FISH, 400 SPECIES OF CORALS, 200 SPECIES OF BIRDS, 30 SPECIES OF WHALES AND DOLPHINS, AND SIX SPECIES OF TURTLES.

UNLIKELY HEROES

A noble-looking dog or a lion with a proud mane are often the types of animal that come to mind as heroes. However, just because a creature might not be beautiful or cuddly, it doesn't mean it can't be heroic.

Rats

Throughout human history, rats have spread plague and other diseases, infested food stores and inhabited sewers, but they do have their uses. Mine Detection Rats have been trained to use their superb sense of smell to detect landmines in Mozambique. Landmines cause 800 deaths and 1,200 injuries worldwide every month, so finding them is a life-saving skill and makes these super-sniffing rodents heroes, not vermin.

Hawks

Pigeons and gulls can cause damage to buildings and they can interrupt sporting events, so birds of prey are sometimes used to scare them away. It wasn't unusual for tennis matches to be halted at the world-famous All England Lawn Tennis Championships, when birds settled on the courts to eat grass seed. Rufus the hawk has been helping to solve the problem for several years.

Honeybees

Many plants rely on insects to pollinate them. The pollination process enables a plant to grow more seeds – so it's an essential part of life on Earth. One of the most important pollinators is the honeybee, which also makes honey and beeswax.

Slugs and Snails

Slimy slugs and snails may not be most people's idea of heroes, but without these animals, and other invertebrates such as worms, our world would be knee-deep in dead plants and animals. The bugs and mini-beasts that devour dead and decaying matter help to keep the planet in good health, and turn organic matter into fertile soil where new plants can grow.

Dung Beetles

A dung beetle collects dung into a ball and lays its eggs inside. When the eggs hatch the beetle larvae feed on the dung. They are just some of the many animals that consume dung, helping to clean the planet by recycling waste.

Dust Mites

Mites are tiny eight-legged relatives of spiders and they live all around us, or on us. Dust mites save us from wallowing in piles of our own dead skin by eating the flakes that fall off our bodies every day as our skin renews.

Mosquitoes

Although female mosquitoes can spread deadly diseases in parts of the world, males have been involved in projects to limit the harm that mosquitoes do. Scientists are able to breed 'sterile males' that mate with females, which prevents the females from reproducing. It's a technique that is helping to control diseases such as malaria and dengue fever.

Cows

Cows helped to destroy an infectious disease called smallpox. In the 18th century, a scientist called Edward Jenner noticed that cowpox, a disease that affected cows, was very similar to smallpox. Many people who contracted smallpox, especially children and babies, were left blind, scarred or died from the illness. Jenner invented the first vaccine by injecting cowpox into people to give them immunity to smallpox. Thanks to Jenner, and the cows, hundreds of millions of lives have been saved worldwide.

QUIZ

1 What did the horse Kerry Gold save farmer Fiona Boyd from?
   a. A pig
   b. A cow
   c. A coyote

2 How long have dogs lived with humans?
   a. At least 32,000 years
   b. Approximately 5,000 years
   c. Less than 500 years

3 What does the name 'orangutan' mean?
   a. Orange ape
   b. Ape of the trees
   c. Person of the forest

4 Which animal helped to destroy smallpox?
   a. Dogs
   b. Cows
   c. Rats

5 Which animals went up in the first 'manned' hot air balloon flight?
   a. A duck, a cockerel and a sheep
   b. A pigeon, a rabbit and a cat
   c. A rat, a dog and a guinea pig

6 What military rank was given to Bert the camel?
   a. Major
   b. Lieutenant
   c. Sergeant

7 What do dogs sniff to tell if children have malaria?
  a. Socks
  b. Fingers
  c. T-shirts

8 What was the name of the conservationist who studied gorillas in Rwanda?
  a. Dian Fossey
  b. Jane Goodall
  c. Biruté Galdikas

9 Which animals help people fish in the seas of Laguna, Brazil?
  a. Orca
  b. Bottlenose dolphins
  c. Whale sharks

10 What kind of animal was Old Blue?
  a. Yangzte river dolphin
  b. Pinta Island tortoise
  c. Chatham Island black robin

11 What name did Koko the gorilla give to her pet kitten?
  a. Whisker Gorilla
  b. All Ball
  c. Baby Koko

12 Which animal predicted the outcome of matches in the 2010 Football World Cup?
  a. Rin Tin Tin the dog
  b. Grumpy Cat
  c. Paul the Octopus

Answers: 1．b, 2．a, 3．c, 4．b, 5．a, 6．c, 7．a, 8．a, 9．b, 10．c, 11．b, 12．c

INDEX

acting 108
animal testing 72
ape 10, 18–19, 46, 47, 106
    chimpanzee 16, 19, 47, 74–75, 76, 78, 79, 107
        Jane Goodall 47
    gorilla 14–15, 44–46, 48–49, 78, 79
        Dian Fossey 44–46
        Ian Redmond 46
    orangutan 18, 19, 47, 78, 79
        Biruté Galdikas 47
avalanches 40
award 23, 72–73, 86, 88

bear 30, 50–51, 52–53, 54–55
    polar bear 53, 55
beaver 13, 30
bird 28, 29, 56, 58, 59, 86, 89, 90, 91, 101
    Arctic tern 90
    bowerbird 77
    canary 72
    chicken 10, 104
    cockatoo 78
    crow 78
    cuckoo 85
    Don Merton 56
    duck 104
    emu 85
    hawk 122
    kingfisher 109
    parrot 12, 17, 59
    penguin 116–118, 119, 120
    pigeon 60, 77, 86–88
    robin 56–57
    vulture 30
bison 30

camel 12, 41, 42
cat 14, 20–22, 23, 24–25, 59, 73, 85, 101, 108
    A Street Cat Named Bob 22
    Grumpy Cat 108
cattle 95, 101, 123
climate change 61, 121
comfort animals 101
communication 14, 16, 31, 34, 49, 64, 71, 76
conservation 26, 46, 47, 49, 79, 82, 112, 114–115
crab 91
crocodile 80, 82

dinosaur 58
dog 8–9, 11, 16, 17, 28, 31, 38–39, 40, 41, 43, 68–69, 70–71, 72, 91, 98–100, 101, 102–103, 106, 108
    Rin Tin Tin 108
    wild dog 16, 28–29
        coyote 16, 28, 30

    dingo 29
    fox 29
    jackal 28, 30
    wolf 11, 26–27, 28, 29, 30, 31, 70, 71
dolphin 34, 36, 37, 61, 78, 121
donkey 42
dugong 121

earthquakes 8, 10
elephant 42, 62–63, 64–65, 66–67, 77
elk 73
endangered 29, 30, 35, 55, 65, 79, 121
extinction 30, 56, 58, 60–61, 79, 113

fame 108
fish 61, 85, 109, 119, 120, 121
    carp 59
    catfish 10
    cod 121
    eel 90
    jawfish 84
    salmon 89
    tuna 121
ferret 101
football 108

goat 101

horse 30, 92–93, 94–95, 96–97, 101
hippopotamus 83
hunting 26, 28, 30, 31, 35, 44, 46, 49, 55, 61, 64, 65, 66, 67, 80, 91, 113, 114

illness 13, 23, 38, 43, 123
insect 16, 85, 106, 107, 122, 123
    ant 16
    bee 10, 17, 77, 122
    butterfly 89, 109
    dragonfly 89
    dung beetle 123
    fly 106, 107
    grasshopper 16
    mosquito 123
    moth 16
intelligence 17, 34, 36, 42, 55, 64, 77, 86
Internet 108

kangaroo 29, 84
koala 84

lemur 79
leopard 83
lion 83, 110–112, 113, 114–115
    Born Free 112, 115
    George Adamson 110, 114–115
lizard 59
llama 41, 73
Louis Leakey 47
lynx 30

marmot 73
meerkat 16
mice 101, 106
migration 65, 89, 90–91
mollusc 121
    slug 122
    snail 122
monkey 17, 18, 78, 106
moose 30

navigation 89, 90

ocean 121
octopus 84, 108
orca 32–33, 34, 35, 36
    Free Willy 32
otter 78, 85
oryx 30

Pablo Picasso 74, 77
painting 66, 74, 77
parenting 11, 31, 49, 80, 84–85, 119
pig 101
pollution 121

rabbit 13, 29, 106
rat 10, 59, 106, 122
reindeer 42
rhino 61

Salvador Dalí 74
science 109
sea lion 77
seahorse 84
senses 11, 16, 29, 43, 71
shark 35, 37, 72, 83, 109, 121
sheep 73, 104–105
sign language 14, 76
snake 10, 59, 77
space travel 106–107
spider 107

tamarin 79
tiger 80–82, 83
toad 59, 61, 84
tools 47, 64, 74, 76, 78
tortoise 60, 107
turtle 91, 121

water bear 107
war 50, 62, 67, 86, 96, 98, 102–103
whale 34–35, 37, 109, 121
wildebeest 89
working animal 42, 94, 96, 97

yak 41

zoo 10, 14, 44, 47, 48, 60, 62, 66, 73, 74, 88, 112, 114